PRAXIS II
MIDDLE SCHOOL
MATHEMATICS 0069, 5169
BOOK AND ONLINE

By: Sharon Wynne, M.S.

XAMonline, INC.
Boston

Library of Congress Cataloging-in-Publication Data

Wynne, Sharon A.
    PRAXIS II Middle School Mathematics 0069, 5169 / Sharon A. Wynne
    ISBN 978-1-60787-399-0
    1. PRAXIS II Middle School Mathematics 0069, 5169
    2. Study Guides
    3. PRAXIS
    4. Teachers' Certification & Licensure
    5. Careers

Disclaimer:

The opinions expressed in this publication are the sole works of XAMonline and were created independently from the National Education Association, Educational Testing Service, or any State Department of Education, National Evaluation Systems or other testing affiliates.

Between the time of publication and printing, state specific standards as well as testing formats and Web site information may change and therefore would not be included in part or in whole within this product. Sample test questions are developed by XAMonline and reflect content similar to that on real tests; however, they are not former test questions. XAMonline assembles content that aligns with state standards but makes no claims nor guarantees teacher candidates a passing score. Numerical scores are determined by testing companies such as NES or ETS and then are compared with individual state standards. A passing score varies from state to state.

Printed in the United States of America   œ-1

PRAXIS II Middle School Mathematics 0069, 5169
ISBN: 978-1-60787-399-0

Table of Contents

Get it now!

To register and access your PRAXIS Mathematics 0069 IntelliGuide™, please visit: XAMonlineIntelliGuide.com/ PRAXISmath69registration or see page viii

DOMAIN I
ARITHMETIC AND BASIC ALGEBRA

DOMAIN II
GEOMETRY AND MEASUREMENT 81

COMPETENCY 2
GEOMETRY AND MEASUREMENT 83

COMPETENCY 4B
DISCRETE MATHEMATICS

DOMAIN V
PROBLEM-SOLVING EXERCISES

SAMPLE TEST

PRAXIS II

MIDDLE SCHOOL MATHEMATICS 0069, 5169 BOOK AND ONLINE

IntelliGuide by XAMonline

Same comprehensive content, new interactive study tools.

XAMonline's new platform transforms the traditional eBook experience into an enhanced study and test practice session through innovative tools such as sticky notes and flashcards. The IntelliGuide™ technology companion is FREE with the purchase of this PRAXIS Middle School Mathematics 0069, 5169 study guide and features:

Easy Navigation: Previously viewed content can be quickly recalled using the "History" tab. This saves time and prevents frustration as you can easily flip back and reference a previous page or topic you recently viewed.

eStickynotes: Allows you to write digital notes anywhere in the study guide, then combine and print out all your notes for easy, "on-the-go" studying.

In-text Search: Where was that topic mentioned? This feature enables you to quickly locate the content you need. Simply type in the search term and the IntelliGuide™ lists the page numbers and highlights where in the page the reference occurs.

eFlashcards: A digital flashcard featuring any combination of words, numbers or symbols provides a learning drill. XAMonline's eFlashcards enable you to study pre-made cards or create your own. The "Hide" function allows you to skip cards you already know, creating a customized study experience. Use them digitally or print them out and study on-the-go.

More Sample Tests Available for Purchase: Additional practice tests assess how much you know and help identify the areas you need to focus on. With up to 125 questions, these exams can be timed to provide you with an experience that mirrors the real test or you can pause the timer as you flip back and reference specific skills or competencies. You can grade the entire completed test or just a portion of it, then determine which content areas and rigor levels you need improvement in.

Get it now!

To register and access your PRAXIS Mathematics 0069 IntelliGuide™, please visit:
XAMonlineIntelliGuide.com/
PRAXISmath69registration

Enter ISBN: 9781607873563

SECTION 1

ABOUT XAMONLINE

XAMonline—A Specialty Teacher Certification Company

Created in 1996, XAMonline was the first company to publish study guides for state-specific teacher certification examinations. Founder Sharon Wynne found it frustrating that materials were not available for teacher certification preparation and decided to create the first single, state-specific guide. XAMonline has grown into a company of over 1,800 contributors and writers and offers over 300 titles for the entire PRAXIS series and every state examination. No matter what state you plan on teaching in, XAMonline has a unique teacher certification study guide just for you.

XAMonline—Value and Innovation

We are committed to providing value and innovation. Our print-on-demand technology allows us to be the first in the market to reflect changes in test standards and user feedback as they occur. Our guides are written by experienced teachers who are experts in their fields. And our content reflects the highest standards of quality. Comprehensive practice tests with varied levels of rigor means that your study experience will closely match the actual in-test experience.

To date, XAMonline has helped nearly 600,000 teachers pass their certification or licensing exams. Our commitment to preparation exceeds simply providing the proper material for study—it extends to helping teachers **gain mastery** of the subject matter, giving them the **tools** to become the most effective classroom leaders possible, and ushering today's students toward a **successful future**.

SECTION 2

ABOUT THIS STUDY GUIDE

Purpose of This Guide

Is there a little voice inside of you saying, "Am I ready?" Our goal is to replace that little voice and remove all doubt with a new voice that says, "I AM READY. **Bring it on!**" by offering the highest quality of teacher certification study guides.

Organization of Content

You will see that while every test may start with overlapping general topics, each is very unique in the skills they wish to test. Only XAMonline presents custom content that analyzes deeper than a title, a subarea, or an objective. Only XAMonline presents content and sample test assessments along with **focus statements**, the deepest-level rationale and interpretation of the skills that are unique to the exam.

Title and field number of test

∞Each exam has its own name and number. XAMonline's guides are written to give you the content you need to know for the specific exam you are taking. You can be confident when you buy our guide that it contains the information you need to study for the specific test you are taking.

Subareas

∞These are the major content categories found on the exam. XAMonline's guides are written to cover all of the subareas found in the test frameworks developed for the exam.

Objectives

∞These are standards that are unique to the exam and represent the main subcategories of the subareas/content categories. XAMonline's guides are written to address every specific objective required to pass the exam.

Focus statements

∞These are examples and interpretations of the objectives. You find them in parenthesis directly following the objective. They provide detailed examples of the range, type, and level of content that appear on the test questions. **Only XAMonline's guides drill down to this level.**

How Do We Compare with Our Competitors?

XAMonline—drills down to the focus statement level.
CliffsNotes and REA—organized at the objective level
Kaplan—provides only links to content
MoMedia—content not specific to the state test

Each subarea is divided into manageable sections that cover the specific skill areas. Explanations are easy to understand and thorough. You'll find that every test answer contains a rejoinder so if you need a refresher or further review after taking the test, you'll know exactly to which section you must return.

How to Use This Book

Our informal polls show that most people begin studying up to eight weeks prior to the test date, so start early. Then ask yourself some questions: How much do you really know? Are you coming to the test straight from your teacher-education program or are you having to review subjects you haven't considered in ten years? Either way, take a **diagnostic or assessment test** first. Also, spend time on sample tests so that you become accustomed to the way the actual test will appear.

This guide comes with an online diagnostic test of 30 questions found online at *www.XAMonline.com*. It is a little boot camp to get you up for the task and reveal things about your compendium of knowledge in general. Although this guide is structured to follow the order of the test, you are not required to study in that order. By finding a time-management and study plan that fits your life you will be more effective. The results of your diagnostic or self-assessment test can be a guide for how to manage your time and point you toward an area that needs more attention.

After taking the diagnostic exam, fill out the **Personalized Study Plan** page at the beginning of each chapter. Review the competencies and skills covered in that chapter and check the boxes that apply to your study needs. If there are sections you already know you can skip, check the "skip it" box. Taking this step will give you a study plan for each chapter.

Week	Activity
8 weeks prior to test	Take a diagnostic test found at www.XAMonline.com
7 weeks prior to test	Build your Personalized Study Plan for each chapter. Check the "skip it" box for sections you feel you are already strong in. ✗ SKIP IT ☐
6-3 weeks prior to test	For each of these four weeks, choose a content area to study. You don't have to go in the order of the book. It may be that you start with the content that needs the most review. Alternately, you may want to ease yourself into plan by starting with the most familiar material.
2 weeks prior to test	Take the sample test, score it, and create a review plan for the final week before the test.
1 week prior to test	Following your plan (which will likely be aligned with the areas that need the most review) go back and study the sections that align with the questions you may have gotten wrong. Then go back and study the sections related to the questions you answered correctly. If need be, create flashcards and drill yourself on any area that you makes you anxious.

SECTION 3
ABOUT THE PRAXIS EXAMS

What Is PRAXIS?

PRAXIS II tests measure the knowledge of specific content areas in K-12 education. The test is a way of insuring that educators are prepared to not only teach in a particular subject area, but also have the necessary teaching skills to be effective. The Educational Testing Service administers the test in most states and has worked with the states to develop the material so that it is appropriate for state standards.

PRAXIS Points

1. The PRAXIS Series comprises more than 140 different tests in over seventy different subject areas.

2. Over 90% of the PRAXIS tests measure subject area knowledge.

3. The purpose of the test is to measure whether the teacher candidate possesses a sufficient level of knowledge and skills to perform job duties effectively and responsibly.

4. Your state sets the acceptable passing score.

5. Any candidate, whether from a traditional teaching-preparation path or an alternative route, can seek to enter the teaching profession by taking a PRAXIS test.

6. PRAXIS tests are updated regularly to ensure current content.

Often **your own state's requirements** determine whether or not you should take any particular test. The most reliable source of information regarding this is either your state's Department of Education or the Educational Testing Service. Either resource should also have a complete list of testing centers and dates. Test dates vary by subject area and not all test dates necessarily include your particular test, so be sure to check carefully.

If you are in a teacher-education program, check with the Education Department or the Certification Officer for specific information for testing and testing timelines. The Certification Office should have most of the information you need.

If you choose an alternative route to certification you can either rely on our Web site at *www.XAMonline.com* or on the resources provided by an alternative certification program. Many states now have specific agencies devoted to alternative certification and there are some national organizations as well:

National Center for Education Information
http://www.ncei.com/Alt-Teacher-Cert.htm

National Associate for Alternative Certification
http://www.alt-teachercert.org/index.asp

Interpreting Test Results

Contrary to what you may have heard, the results of a PRAXIS test are not based on time. More accurately, you will be scored on the raw number of points you earn in relation to the raw number of points available. Each question is worth one raw point. It is likely to your benefit to complete as many questions in the time allotted, but it will not necessarily work to your advantage if you hurry through the test.

Follow the guidelines provided by ETS for interpreting your score. The web site offers a sample test score sheet and clearly explains how the scores are scaled and what to expect if you have an essay portion on your test.

Scores are usually available by phone within a month of the test date and scores will be sent to your chosen institution(s) within six weeks. Additionally, ETS now makes online, downloadable reports available for 45 days from the reporting date.

It is **critical** that you be aware of your own state's passing score. Your raw score may qualify you to teach in some states, but not all. ETS administers the test and assigns a score, but the states make their own interpretations and, in some cases, consider combined scores if you are testing in more than one area.

What's on the Test?

PRAXIS tests vary from subject to subject and sometimes even within subject area. For PRAXIS Middle School Mathematics 0069, 5169 the test lasts for 2 hours and consists of two sections. Part A consists of approximately 40 multiple-choice questions and Part B consists of 3 short constructed response questions. The use of graphing calculators is permitted for this test but calculators with QWERTY keyboards are not. The breakdown of the questions is as follows:

Category	Approximate Number of Questions	Approximate Percentage of the test	Percentage of Total Score
I: Arithmetic and Basic Algebra	12	20%	Multiple choice section equates to 67% of the total score
II: Geometry and Measurement	10	17%	

Table continued on next page

Category	Approximate Number of Questions	Approximate Percentage of the test	Percentage of Total Score
III: Functions and Their Graphs	8	13%	
IV: Data, Probability, and Statistical Concepts; Discrete Mathematics	10	17%	
V: Problem-Solving Exercises	3 (constructed response)	33%	33%

The following process categories are distributed throughout the test questions in each category:

- Mathematical Problem Solving
- Mathematical Reasoning and Proof
- Mathematical Connections
- Mathematical Representation
- Use of Technology

This chart can be used to build a study plan. Twenty percent may seem like a lot of time to spend on Arithmetic and Basic Algebra, but when you consider that amounts to about 1 out of 5 multiple choice questions, it might change your perspective.

Question Types

You're probably thinking, enough already, I want to study! Indulge us a little longer while we explain that there is actually more than one type of multiple-choice question. You can thank us later after you realize how well prepared you are for your exam.

1. **Complete the Statement.** The name says it all. In this question type you'll be asked to choose the correct completion of a given statement. For example:

> **The Dolch Basic Sight Words consist of a relatively short list of words that children should be able to:**
>
> A. Sound out
>
> B. Know the meaning of
>
> C. Recognize on sight
>
> D. Use in a sentence

The correct answer is C. In order to check your answer, test out the statement by adding the choices to the end of it.

2. Which of the Following. One way to test your answer choice for this type of question is to replace the phrase "which of the following" with your selection. Use this example:

Which of the following words is one of the twelve most frequently used in children's reading texts:

A. There

B. This

C. The

D. An

Don't look! Test your answer. _____ is one of the twelve most frequently used in children's reading texts. Did you guess C? Then you guessed correctly.

3. Roman Numeral Choices. This question type is used when there is more than one possible correct answer. For example:

Which of the following two arguments accurately supports the use of cooperative learning as an effective method of instruction?

I.   Cooperative learning groups facilitate healthy competition between individuals in the group.

II. Cooperative learning groups allow academic achievers to carry or cover for academic underachievers.

III. Cooperative learning groups make each student in the group accountable for the success of the group.

IV. Cooperative learning groups make it possible for students to reward other group members for achieving.

A. I and II

B. II and III

C. I and III

D. III and IV

Notice that the question states there are **two** possible answers. It's best to read all the possibilities first before looking at the answer choices. In this case, the correct answer is D.

4. **Negative Questions.** This type of question contains words such as "not," "least," and "except." Each correct answer will be the statement that does **not** fit the situation described in the question. Such as:

Multicultural education is not

A. An idea or concept

B. A "tack-on" to the school curriculum

C. An educational reform movement

D. A process

Think to yourself that the statement could be anything but the correct answer. This question form is more open to interpretation than other types, so read carefully and don't forget that you're answering a negative statement.

5. **Questions that Include Graphs, Tables, or Reading Passages.** As always, read the question carefully. It likely asks for a very specific answer and not a broad interpretation of the visual. Here is a simple (though not statistically accurate) example of a graph question:

In the following graph in how many years did more men take the NYSTCE exam than women?

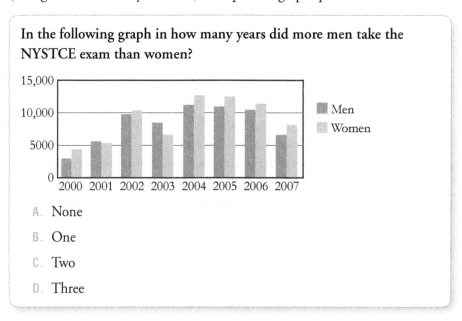

A. None

B. One

C. Two

D. Three

It may help you to simply circle the two years that answer the question. Make sure you've read the question thoroughly and once you've made your determination, double check your work. The correct answer is C.

SECTION 4
HELPFUL HINTS

Study Tips

1. **You are what you eat.** Certain foods aid the learning process by releasing natural memory enhancers called CCKs (cholecystokinin) composed of tryptophan, choline, and phenylalanine. All of these chemicals enhance the neurotransmitters associated with memory and certain foods release memory enhancing chemicals. A light meal or snacks of one of the following foods fall into this category:

   - Milk
   - Rice
   - Eggs
   - Fish
   - Nuts and seeds
   - Oats
   - Turkey

   The better the connections, the more you comprehend!

2. **See the forest for the trees.** In other words, get the concept before you look at the details. One way to do this is to take notes as you read, paraphrasing or summarizing in your own words. Putting the concept in terms that are comfortable and familiar may increase retention.

3. **Question authority.** Ask why, why, why? Pull apart written material paragraph by paragraph and don't forget the captions under the illustrations. For example, if a heading reads *Stream Erosion* put it in the form of a question (Why do streams erode? What is stream erosion?) then find the answer within the material. If you train your mind to think in this manner you will learn more and prepare yourself for answering test questions.

4. **Play mind games.** Using your brain for reading or puzzles keeps it flexible. Even with a limited amount of time your brain can take in data (much like a computer) and store it for later use. In ten minutes you can: read two paragraphs (at least), quiz yourself with flash cards, or review notes. Even if you don't fully understand something on the first pass, your mind stores it for recall, which is why frequent reading or review increases chances of retention and comprehension.

5. **The pen is mightier than the sword.** Learn to take great notes. A by-product of our modern culture is that we have grown accustomed to getting our information in short doses. We've subconsciously trained ourselves to assimilate information into neat little packages. Messy notes fragment the flow of information. Your notes can be much clearer with proper formatting. ***The Cornell Method*** is one such format. This method was popularized in *How to Study in College*, Ninth Edition, by Walter Pauk. You can benefit from the method without purchasing an additional book by simply looking up the method online. Below is a sample of how *The Cornell Method* can be adapted for use with this guide.

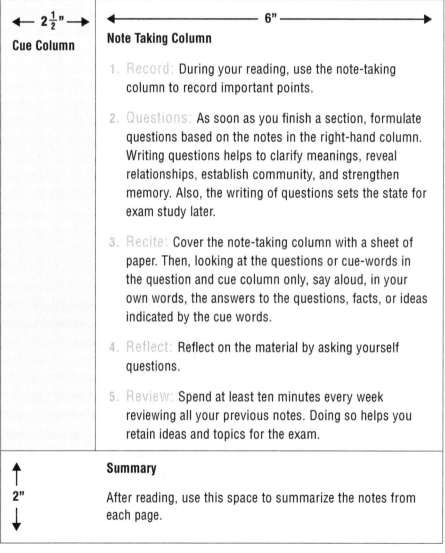

Adapted from How to Study in College, Ninth Edition, by Walter Pauk, ©2008 Wadsworth

6. **Place yourself in exile and set the mood.** Set aside a particular place and time to study that best suits your personal needs and biorhythms. If you're a night person, burn the midnight oil. If you're a morning person set yourself up with some coffee and get to it. Make your study time and place as free from distraction as possible and surround yourself with what you need, be it silence or music. Studies have shown that music can aid in concentration, absorption, and retrieval of information. Not all music, though. Classical music is said to work best

7. **Get pointed in the right direction.** Use arrows to point to important passages or pieces of information. It's easier to read than a page full of yellow highlights. Highlighting can be used sparingly, but add an arrow to the margin to call attention to it.

8. **Check your budget.** You should at least review all the content material before your test, but allocate the most amount of time to the areas that need the most refreshing. It sounds obvious, but it's easy to forget. You can use the study rubric above to balance your study budget.

> *The proctor will write the start time where it can be seen and then, later, provide the time remaining, typically fifteen minutes before the end of the test.*

Testing Tips

1. **Get smart, play dumb.** Sometimes a question is just a question. No one is out to trick you, so don't assume that the test writer is looking for something other than what was asked. Stick to the question as written and don't overanalyze.

2. **Do a double take.** Read test questions and answer choices at least twice because it's easy to miss something, to transpose a word or some letters. If you have no idea what the correct answer is, skip it and come back later if there's time. If you're still clueless, it's okay to guess. Remember, you're scored on the number of questions you answer correctly and you're not penalized for wrong answers. The worst case scenario is that you miss a point from a good guess.

3. **Turn it on its ear.** The syntax of a question can often provide a clue, so make things interesting and turn the question into a statement to see if it changes the meaning or relates better (or worse) to the answer choices.

4. **Get out your magnifying glass.** Look for hidden clues in the questions because it's difficult to write a multiple-choice question without giving away part of the answer in the options presented. In most questions you can readily eliminate one or two potential answers, increasing your chances of answering correctly to 50/50, which will help out if you've skipped a question and gone back to it (see tip #2).

5. Call it intuition. Often your first instinct is correct. If you've been study-ing the content you've likely absorbed something and have subconsciously retained the knowledge. On questions you're not sure about trust your instincts because a first impression is usually correct.

6. Graffiti. Sometimes it's a good idea to mark your answers directly on the test booklet and go back to fill in the optical scan sheet later. You don't get extra points for perfectly blackened ovals. If you choose to manage your test this way, be sure not to mismark your answers when you transcribe to the scan sheet.

7. Become a clock-watcher. You have a set amount of time to answer the questions. Don't get bogged down laboring over a question you're not sure about when there are ten others you could answer more readily. If you choose to follow the advice of tip #6, be sure you leave time near the end to go back and fill in the scan sheet.

Do the Drill

No matter how prepared you feel it's sometimes a good idea to apply Murphy's Law. So the following tips might seem silly, mundane, or obvious, but we're including them anyway.

1. Remember, you are what you eat, so bring a snack. Choose from the list of energizing foods that appear earlier in the introduction.

2. You're not too sexy for your test. Wear comfortable clothes. You'll be distracted if your belt is too tight or if you're too cold or too hot.

3. Lie to yourself. Even if you think you're a prompt person, pretend you're not and leave plenty of time to get to the testing center. Map it out ahead of time and do a dry run if you have to. There's no need to add road rage to your list of anxieties.

4. Bring sharp number 2 pencils. It may seem impossible to forget this need from your school days, but you might. And make sure the erasers are intact, too.

5. No ticket, no test. Bring your admission ticket as well as **two** forms of identification, including one with a picture and signature. You will not be admitted to the test without these things.

6. You can't take it with you. Leave any study aids, dictionaries, note-books, computers, and the like at home. Certain tests **do** allow a scientific or four-function calculator, so check ahead of time to see if your test does.

7. Prepare for the desert. Any time spent on a bathroom break **cannot** be made up later, so use your judgment on the amount you eat or drink.

8. Quiet, Please! Keeping your own time is a good idea, but not with a timepiece that has a loud ticker. If you use a watch, take it off and place it nearby but not so that it distracts you. And **silence your cell phone**.

To the best of our ability, we have compiled the content you need to know in this book and in the accompanying online resources. The rest is up to you. You can use the study and testing tips or you can follow your own methods. Either way, you can be confident that there aren't any missing pieces of information and there shouldn't be any surprises in the content on the test.

If you have questions about test fees, registration, electronic testing, or other content verification issues please visit *www.ets.org*.

Good luck!

Sharon Wynne
Founder, XAMonline

DOMAIN I
ARITHMETIC AND BASIC ALGEBRA

PERSONALIZED STUDY PLAN

KNOWN MATERIAL/ SKIP IT

PAGE	COMPETENCY AND SKILL	
3	**1: Arithmetic and basic algebra**	☐
	1.1: Add, subtract, multiply and divide rational numbers expressed in various forms	☐
	1.2: Apply the order of operations	☐
	1.3: Identify the properties of the basic operations on the standard number systems	☐
	1.4: Identify an inverse and the additive and multiplicative inverses of a number	☐
	1.5: Use numbers in a way that is most appropriate in the context of a problem	☐
	1.6: Order any finite set of real numbers and recognize equivalent forms of a number	☐
	1.7: Classify a number as rational, irrational, real, or complex	☐
	1.8: Estimate values of expressions involving decimals, exponents, and radicals	☐
	1.9: Find powers and roots	☐
	1.10: Given newly defined operations on a number system, determine whether the closure, commutative, associative, or distributive properties hold	☐
	1.11: Demonstrate an understanding of concepts associated with counting numbers	☐
	1.12: Interpret and apply the concepts of ratio, proportion, and percent in appropriate situations	☐
	1.13: Recognize the reasonableness of results within the context of a given problem; using estimation, test the reasonableness of results	☐
	1.14: Work with algebraic expressions, formulas, and equations; add, subtract, and multiply polynomials; divide polynomials	☐
	1.15: Add, subtract, multiply, and divide algebraic fractions	☐
	1.16: Perform standard algebraic operations involving complex numbers	☐
	1.17: Perform standard operations involving radicals and exponents, including fractional and negative exponents	☐
	1.18: Determine the equations of lines; recognize and use the basic forms of the equations for a straight line	☐
	1.19: Solve and graph linear equations and inequalities in one or two variables; solve and graph systems of linear equations and inequalities in two variables	☐
	1.20: Solve and graph nonlinear algebraic equations	☐
	1.21: Solve equations and inequalities involving absolute values	☐
	1.22: Solve problems that involve quadratic equations using a variety of methods	☐

COMPETENCY 1
ARITHMETIC AND BASIC ALGEBRA

Addition can be indicated by the expressions *sum, greater than, and, more than, increased by,* and *added to.* Subtraction can be expressed by difference, fewer than, minus, less than, decreased by, and subtracted from. Addition and subtraction can be demonstrated with symbols, as shown in the following example.

ψψψ𝜁𝜁𝜁𝜁

$3 + 4 = 7$

$7 - 3 = 4$

Multiplication is shown by *product, times, multiplied by,* and *twice.* Multiplication can be shown using arrays. For instance, 3×4 may be expressed as 3 rows of 4 each:

☐ ☐ ☐ ☐
☐ ☐ ☐ ☐
☐ ☐ ☐ ☐

Division is expressed by *quotient, divided by,* and *ratio.* As with multiplication, division can be demonstrated with an array. For example, $12 \div 2$ can be shown by placing 12 objects in 2 rows:

☐ ☐ ☐ ☐ ☐ ☐
☐ ☐ ☐ ☐ ☐ ☐

So it is obvious that the answer is 6.

Examples:

7 added to a number	$n + 7$
a number decreased by 8	$n - 8$
12 times a number divided by 7	$12n \div 7$
28 less than a number	$n - 28$
the ratio of a number to 55	$\frac{n}{55}$
4 times the sum of a number and 21	$4(n + 21)$

Mathematical operations can be shown using manipulatives or drawings. Fractions can be clarified using pattern blocks, fraction bars, or paper folding.

Generally, when applying these operations to rational numbers, the same rules apply. Any rational number can be expressed as a fraction with a finite numerator and finite denominator. Even whole numbers (such as 0, 1, 2, etc.) can be expressed as fractions ($\frac{0}{1}, \frac{1}{1}, \frac{2}{1}$, etc.).

For addition or subtraction of rational fractions, the numerators can be added or subtracted when the denominators are equal. For instance:

$$\frac{1}{5} + \frac{2}{5} = \frac{1+2}{5} = \frac{3}{5}$$

If the denominators are different, find the least common denominator and convert the fractions appropriately. For instance:

$$\frac{2}{3} - \frac{1}{9} = \left(\frac{2}{3} \times \frac{3}{3}\right) - \frac{1}{9} = \frac{6}{9} - \frac{1}{9} = \frac{6-1}{9} = \frac{5}{9}$$

For multiplication of two fractions, the result is simply the product of the numerators over the product of the denominators.

$$\frac{3}{4} \times \frac{1}{7} = \frac{3 \times 1}{4 \times 7} = \frac{3}{28}$$

Division involves multiplication by the reciprocal of the divisor. Thus, for example:

$$\frac{4}{5} \div \frac{7}{2} = \frac{4}{5} \times \frac{2}{7} = \frac{8}{35}$$

When performing operations on decimals, one option is to convert the decimals to fractions (this may be helpful when multiplying long decimals that can be represented as simple fractions). A calculator can be used for performing operations on whole numbers or decimals, but be aware that the answer may not be accurate. The limitations of calculator technology mean that, for instance, the answer to $8 \div 9$ may show up as 0.888888889, but the actual result is $0.\overline{8}$ (an infinite number of eights after the decimal point).

SKILL 1.2 Apply the order of operations

The order of operations must be followed when evaluating algebraic expressions. The mnemonic PEMDAS (Please Excuse My Dear Aunt Sally) can help you to remember the order of the following steps.

1. Simplify inside grouping characters such as parentheses (P for Please), brackets, radicals, fraction bars, and so on.

2. Multiply out expressions with exponents (E for Excuse).

3. Do multiplication (M for My) or division (D for Dear) from left to right.

4. Do addition (A for Aunt) or subtraction (S for Sally) from left to right.

Here are some samples of simplifying expressions with exponents:

SIMPLIFIED EXPRESSIONS WITH EXPONENTS			
$(-2)^3$	=	-8	$-2^3 = -8$
$(-2)^4$	=	16	$-2^4 = -16$ Note change of sign.
$(\frac{2}{3})^3$	=	$\frac{8}{27}$	
5^0	=	1	
4^{-1}	=	$\frac{1}{4}$	

SKILL 1.3 Identify the properties of the basic operations on the standard number systems (e.g., closure, commutativity, associativity, distributivity)

Real numbers exhibit the following addition and multiplication properties, where a, b, and c are real numbers.

Note: Multiplication is implied when there is no symbol between two variables. Thus,

$a \times b$ can be written ab.

Multiplication can also be indicated by a raised dot · as in ($a \cdot b$).

Closure

The Closure Property for Whole-Number Addition states that the sum of any two whole numbers is a whole number.

Example: Since 2 and 5 are both whole numbers, 7 is also a whole number.
The Closure Property for Whole-Number Multiplication states that the product of any two whole numbers is a whole number.

Example: Since 3 and 4 are both whole numbers, 12 is also a whole number. The sum or product of two whole numbers is a whole number.

For any set of numbers to be closed under an operation, the result of the operation on two numbers in the set must also be included within that set.

Commutativity
$a + b = b + a$
Example: $5 + -8 = -8 + 5 = -3$

$ab = ba$
Example: $-2 \times 6 = 6 \times -2 = -12$

The order of the addends or factors does not affect the sum or product.

Associativity
$(a + b) + c = a + (b + c)$
Example: $(-2 + 7) + 5 = -2 + (7 + 5)$
  $5 + 5 = -2 + 12 = 10$

$(ab)c = a(bc)$
Example: $(3 \times -7) \times 5 = 3 \times (-7 \times 5)$
  $-21 \times 5 = 3 \times -35 = -105$

The grouping of the addends or factors does not affect the sum or product.

Distributivity
$a(b + c) = ab + ac$
Example: $6 \times (-4 + 9) = (6 \times -4) + (6 \times 9)$
  $6 \times 5 = -24 + 54 = 30$

To multiply a sum by a number, multiply each addend by the number, and then add the products.

Additive Identity (Property of Zero)
$a + 0 = a$
Example: $17 + 0 = 17$

The sum of any number and zero is that number.

Multiplicative Identity (Property of One)

a × 1 = a
Example: -34 × 1 = -34

The product of any number and 1 is that number.

SKILL **Identify an inverse and the additive and multiplicative inverses of a**
1.4 **number**

An INVERSE of a number can have one of two senses: an additive inverse, which involves negation of the number (or, equivalently, multiplication by -1), or a multiplicative inverse, which involves the reciprocal of the number (or, equivalently, applying an exponent of -1 to the number). Thus, for any number a, the additive inverse is -a, and the multiplicative inverse is $\frac{1}{a}$. Inverse numbers obey the following properties.

> **INVERSE:** the inverse of a number can have one of two senses: an additive inverse, which involves negation of the number, or a multiplicative inverse, which involves the reciprocal of the number

Additive Inverse (Property of Opposites)

a + -a = 0
Example: 25 + -25 = 0

The sum of any number and its opposite is zero.

Multiplicative Inverse (Property of Reciprocals)

$a \times \frac{1}{a} = 1$
Example: $5 \times \frac{1}{5} = 1$

The product of any number and its reciprocal is 1.

These inverse operations are used when solving equations.

SKILL **Use numbers in a way that is most appropriate in the context of a**
1.5 **problem**

See Skill 1.13 for more on estimation and approximation

Example: The 7 a.m. train from the suburbs into the city has five cars that each carry a maximum of 46 passengers. All of the cars are nearly full of passengers. Each passenger has about the same amount of cash. Approximately how much cash is on the train?

   A. $500

   B. $10,000

   C. $100,000

   D. $150,000

The most logical answer is b, because the train is carrying about 230 people. Most of the passengers probably have $20 to $50 each.

SKILL 1.6 **Order any finite set of real numbers and recognize equivalent forms of a number**

Property of Denseness

Between any pair of rational numbers, there is at least one rational number. The set of natural numbers is not dense because between two consecutive natural numbers, there may not exist another natural number.

Example: Between 7.6 and 7.7, there is the rational number 7.65 in the set of real numbers. Between 3 and 4, there exists no other natural number.

If we compare numbers in various forms, we see that:

The integer $4 = \frac{8}{2}$ (fraction) = 4.0 (decimal) = 400% (percent).

From this, you should be able to determine that fractions, decimals, and percents can be used interchangeably within problems.

- To change a percent into a decimal, move the decimal point two places to the left and drop off the percent sign.

- To change a decimal into a percent, move the decimal two places to the right and add on a percent sign.

- To change a fraction into a decimal, divide the numerator by the denominator.

- To change a decimal number into an equivalent fraction, write the decimal part of the number as the fraction's numerator. As the fraction's denominator, use the place value of the last column of the decimal. Reduce the resulting fraction as far as possible.

Example: J.C. Nickels has Hunch jeans for sale at 25% off the usual price of $36.00. Shears and Roadster have the same jeans for sale at 30% off their regular price of $40. Find the cheaper price.

$\frac{1}{4}$ = .25, so .25(36) = $9.00 off; $36 − 9 = $27 sale price

30% = .30, so .30(40) = $12 off; $40 − 12 = $28 sale price

The price at J.C Nickels is $1 lower.

To convert a fraction to a decimal, as we did in the example above, simply divide the numerator (top) by the denominator (bottom). Use long division if necessary.

If a decimal has a fixed number of digits, the decimal is said to be a TERMINATING DECIMAL. To write such a decimal as a fraction, first determine what place value the digit farthest to the right has (for example: tenths, hundredths, thousandths, ten-thousandths, hundred-thousandths, etc.). Then drop the decimal point and place the string of digits over the number given by the place value.

If a decimal continues forever by repeating a string of digits, the decimal is said to be a REPEATING DECIMAL. To write a repeating decimal as a fraction, follow these steps:

1. Let x = the repeating decimal (Ex. x = .716716716...)

2. Multiply x by the multiple of 10 that will move the decimal just to the right of the repeating block of digits (Ex. $1000x$ = 716.716716...)

3. Subtract the first equation from the second (Ex. $1000x − x$ = 716.716716... − .716716...)

4. Simplify and solve this equation. The repeating block of digits will subtract out (Ex. $999x$ = 716 so $x = \frac{716}{999}$)

5. The solution will be the fraction for the repeating decimal

> **TERMINATING DECIMAL:** a decimal that has a fixed number of digits

> **REPEATING DECIMAL:** a decimal that continues forever by repeating a string of digits

COMMON EQUIVALENTS				
$\frac{1}{2}$	=	0.5	=	50%
$\frac{1}{3}$	=	0.33	=	$33\frac{1}{3}$%
$\frac{1}{4}$	=	0.25	=	25%
$\frac{1}{5}$	=	0.2	=	20%
$\frac{1}{6}$	=	0.17	=	$16\frac{2}{3}$%

Table continued on next page

$\frac{1}{8}$	=	0.125	=	$12\frac{1}{2}\%$
$\frac{1}{10}$	=	0.1	=	10%
$\frac{2}{3}$	=	0.67	=	$66\frac{2}{3}\%$
$\frac{5}{6}$	=	0.83	=	$83\frac{1}{3}\%$
$\frac{3}{8}$	=	0.375	=	$37\frac{1}{2}\%$
$\frac{5}{8}$	=	0.625	=	$62\frac{1}{2}\%$
$\frac{7}{8}$	=	0.875	=	$87\frac{1}{2}\%$
1	=	1.0	=	100%

Decimal values have been rounded off to the hundredths place.

SKILL 1.7 Classify a number as rational, irrational, real, or complex

Real Numbers

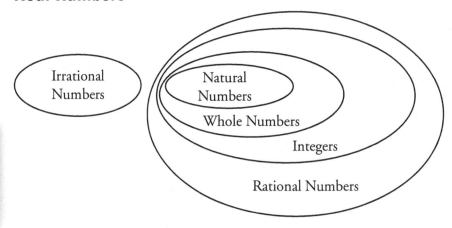

REAL NUMBERS:
numbers that can be
shown by an infinite
decimal representation
such as 3.286275347

*For a listing of
mathematical
symbols, go to:*

*http://en.wikipedia.org/
wiki/Math_symbols*

REAL NUMBERS are denoted by ℝ and are numbers that can be shown by an infinite decimal representation such as 3.286275347. . . . The real numbers include rational numbers such as 42 and –23/129, and irrational numbers such as $\sqrt{2}$ and π, which can be represented as points along an infinite number line. Real numbers are also known as "the unique complete Archimedean ordered field." Real numbers are differentiated from imaginary numbers.

Properties of Real Numbers

Real numbers have the two primary properties of being an ordered field and having the least upper-bound property. The First Property says that real numbers comprise an ordered field (with addition and multiplication, as well as division by nonzero numbers) that can be ordered on a number line in a way that works with addition and multiplication.

Example: O is an ordered field if the order satisfies the following properties:
　If $a \leq b$, then $a + c \leq b + c$.
　If $0 \leq a$ and $0 \leq b$, then $0 \leq ab$.

It then follows that for every a, b, c, d in O:
　Either $-a \leq 0 \leq a$ or $a \leq 0 \leq -a$.

We can add inequalities:
Example: If $a \leq b$ and $c \leq d$, then $a + c \leq b + d$.
　We are allowed to "multiply inequalities with positive elements": If $a \leq b$ and $0 \leq c$, then $ac \leq bc$.

The Second Property of real numbers says that if a nonempty set of real numbers has an upper bound (e.g., $\leq$, or less than or equal to), then it has a least upper bound. These two properties together define the real numbers completely and allow their other properties to be inferred: Every polynomial of odd degree with real coefficients has a real root. If you add the square root of -1 to the real numbers, you have a complex number, and the result is algebraically closed.

Real numbers are classified as follows:

- Natural numbers, denoted by $\mathbb{N}$: the counting numbers 1,2,3,...

- Whole numbers: the counting numbers along with zero 0,1,2...

- Integers, denoted by $\mathbb{Z}$: the counting numbers, their opposites, and zero ..., $-1,0,1,...$

- Rationals, denoted by $\mathbb{Q}$: all of the fractions that can be formed from the whole numbers. Zero cannot be the denominator. In decimal form, these numbers will either be terminating or repeating decimals. Simplify square roots to determine if the number can be written as a fraction.

- Irrationals: real numbers that cannot be written as a fraction. The decimal forms of these numbers are neither terminating nor repeating. Examples: π, e, $\sqrt{2}$, and so on

- Real numbers, denoted by $\mathbb{R}$: the set of numbers obtained by combining the rationals and the irrationals.

Complex Numbers

Complex numbers, i.e. numbers that involve i or $\sqrt{-1}$, are not real numbers. Complex numbers generally include all numbers that have a real and an imaginary part. The imaginary part of a complex number is some real number (which may be zero) multiplied by i (or $\sqrt{-1}$). Thus, the general form of a complex number is $a + bi$, where a and b are real numbers. Since it is possible that $b = 0$, the set of complex numbers includes the set of real numbers.

Properties of complex numbers

- We just reviewed real numbers; real numbers can be ordered, but complex numbers cannot be ordered

- When i appears in a fraction, the fraction is usually simplified so that i is not in the denominator

The complex plane and complex numbers as ordered pairs

Every complex number $a + bi$ can be shown as a pair of two real numbers. For the real part a and the imaginary part bi, b is also real.

Example: 3i has a real part 0 and an imaginary part 3, and 4 has a real part 4 and an imaginary part 0. As another way of writing complex numbers, we can write them as ordered pairs:

Complex Number	Ordered Pair
$3 + 2i$	$(3, 2)$
$\sqrt{3}, \sqrt{3}i$	$(\sqrt{3}, \sqrt{3})$
$7i$	$(0, 7)$
$\dfrac{6 + 2i}{7}$	$\left(\dfrac{6}{7}, \dfrac{2}{7}\right)$

When i^2 appears in a problem, it can be replaced by -1, since $i^2 = -1$. How do we turn 1 into -1? We can work with what is called a complex plane and visually see how it can be done. Instead of an *x-axis* and a *y-axis*, we have an axis that represents the real dimension and an axis that represents the imaginary dimension. If we rotate x 180° in a counterclockwise direction, we can change 1 into -1, which is the same as multiplying 1 by i^2.

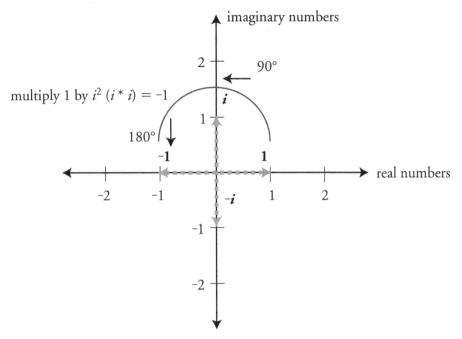

We could also rotate clockwise 180° to turn -1 into 1. This is a multiplication by i^2.

If we multiply $-i$ by $-i$, we get -1. Therefore, there are two square roots of -1: i and $-i$.

SKILL Estimate values of expressions involving decimals, exponents,
1.8 and radicals

Exponents

The EXPONENT FORM is a shortcut method to write repeated multiplication. The BASE is the factor being multiplied. The EXPONENT tells how many times that number is multiplied by itself.

Example: 3^4 is $3 \times 3 \times 3 \times 3 = 81$,
    where 3 is the base and 4 is the exponent.
    x^2 is read as "x squared."
    y^3 is read as "y cubed."

EXPONENT FORM: a shortcut method to write repeated multiplication

BASE: the factor being multiplied

EXPONENT: tells how many times that number is multiplied by itself

$a^1 = a$ for all values of a; thus, $17^1 = 17$.

$b^0 = 1$ for all values of b; thus, $24^0 = 1$.

When 10 is raised to any power, the exponent tells the number of zeros in the product.

Example: $10^7 = 10,000,000$

RULES OF EXPONENTS	
$a^x a^y = a^{x+y}$	*Example: $(3^4)(3^5) = 3^9$*
$a^x b^x = (ab)^x$	*Example: $(4^2)(5^2) = 20^2$*
$(a^x)^y = a^{xy}$	*Example: $(2^3)^2 = 2^6$*
$a^x \div a^y = a^{x-y}$	*Example: $2^5 \div 2^3 = 2^2$*
$a^{-x} = \dfrac{1}{a^x}$	*Example: $2^{-2} = \dfrac{1}{2^2}$*

Scientific Notation

SCIENTIFIC NOTATION is a more convenient method for writing very large and very small numbers. It employs two factors: a number between -10 and 10 and a power of 10. This notation is a shorthand way to express large numbers, such as the weight of 100 freight cars in kilograms, or small numbers, such as the weight of an atom in grams.

SCIENTIFIC NOTATION: a method for writing very large and very small numbers

COMMON MEASURES USING SCIENTIFIC NOTATION			
10^n	=	$(10)^n$ Ten multiplied by itself n times	
10^6	=	1,000,000	(mega)
10^3	=	$10 \times 10 \times 10 = 1000$	(kilo)
10^2	=	$10 \times 10 = 100$	(hecto)
10^1	=	10	(deca)
10^0	=	Any nonzero number raised to the zero power is 1	
10^{-1}	=	$\dfrac{1}{10}$	(deci)
10^{-2}	=	$\dfrac{1}{100}$	(centi)

Table continued on next page

10^{-3}	$=$	$\dfrac{1}{1000}$	(milli)
10^{-6}	$=$	$\dfrac{1}{1,000,000}$	(micro)

To write a number in scientific notation, convert the number to a form of $b \times 10^n$, where $-10 < b < 10$ and n is an integer.

Example: 356.73 can be written in various forms:

$$
\begin{aligned}
356.73 &= 3567.3 \times 10^{-1} \quad (1) \\
&= 35673 \times 10^{-2} \quad (2) \\
&= 35.673 \times 10^{1} \quad (3) \\
&= 3.5673 \times 10^{2} \quad (4) \\
&= 0.35673 \times 10^{3} \quad (5)
\end{aligned}
$$

Only (4) is written in the proper scientific notation format, because 3.5673 (*b*) is a number between -10 and 10.

To change a number into scientific notation, move the decimal point so that only one number from 1 to 9 is in front of the decimal point. Remove any trailing zeros. Multiply this number times 10 to a power. The power is the number of positions that the decimal point is moved. The power is negative if the original number is a decimal number between 1 and -1. Otherwise, the power is positive.

Example: Change into scientific notation:

4,380,000,000	Move decimal behind the 4.
4.38	Remove trailing zeros.
$4.38 \times 10^{?}$	Count the positions that the decimal point has moved.
4.38×10^{9}	This is the answer.
-.0000407	Move the decimal behind the 4.
-4.07	Count the positions that the decimal point has moved.
-4.07×10^{-5}	Note the negative exponent.

If a number is already in scientific notation, it can be changed back into the regular decimal form. If the exponent on the number 10 is negative, move the decimal point to the left. If the exponent on the number 10 is positive, move the decimal point to the right that number of places.

Example: Write 46,368,000 in scientific notation.

1. Introduce a decimal point: 46,368,000 = 46,368,000.0.

2. Move the decimal place to the left until only one nonzero digit is in front of it—in this case, between the 4 and the 6.

3. Count the number of digits that the decimal point has moved—in this case, 7. This is the nth power of ten, and it is positive because the decimal point moved left.

4. Therefore, $46,368,000 = 4.6368 \times 10^7$.

Example: Write 0.00397 in scientific notation.

1. The decimal point is already in place.

2. Move the decimal point to the *right* until there is only one nonzero digit in front of it—in this case, between the 3 and the 9.

3. Count the number of digits the decimal point moved—in this case, 3. This is the nth power of ten, and it is *negative* because the decimal point moved right.

4. Therefore, $0.00397 = 3.97 \times 10^{-3}$.

Example: Evaluate $\dfrac{3.22 \times 10^{-3} \times 736}{0.00736 \times 32.2 \times 10^{-6}}$

Since we have a mixture of large and small numbers, convert each number to scientific notation:

$736 = 7.36 \times 10^2$

$0.00736 = 7.36 \times 10^{-3}$

$32.2 \times 10^{-6} = 3.22 \times 10^{-5}$

Thus, we have

$$\frac{3.22 \times 10^{-3} \times 7.36 \times 10^2}{7.36 \times 10^{-3} \times 3.22 \times 10^{-5}}$$
$$= \frac{3.22 \times 7.36 \times 10^{-3} \times 10^2}{7.36 \times 3.22 \times 10^{-3} \times 10^{-5}}$$
$$= \frac{3.22 \times 7.36}{7.36 \times 3.22} \times \frac{10^{-1}}{10^{-8}}$$
$$= \frac{3.22 \times 7.36}{3.22 \times 7.36} \times 10^{-1} \times 10^8$$
$$= \frac{23.6992}{23.6992} \times 10^7$$
$$= 1 \times 10^7 = 10,000,000$$

Example: Change back into decimal form:

3.448×10^{-2}	Move the decimal point 2 places left, since the exponent is negative.
.003448	This is the answer.
6×10^4	Move the decimal point 4 places right, since the exponent is positive.
60,000	This is the answer.

To add or subtract in scientific notation, the exponents must be the same. Then add the decimal portions, keeping the power of 10 the same. Then move the decimal point and adjust the exponent to keep the number in front of the decimal point from 1 to 9.

Example:

$$
\begin{array}{l}
6.22 \times 10^3 \\
\underline{+\ 7.48 \times 10^3} \qquad \text{Add these as is.} \\
13.70 \times 10^3 \qquad \text{Now move decimal 1 more place to the left and} \\
1.37 \times 10^4 \qquad \text{add 1 more exponent.}
\end{array}
$$

To multiply or divide in scientific notation, multiply or divide the decimal part of the numbers. In multiplication, add the exponents of 10. In division, subtract the exponents of 10. Then move the decimal point and adjust the exponent to keep the number in front of the decimal point from 1 to 9.

Example:

$(5.2 \times 10^5)(3.5 \times 10^2)$ Mulitply 5.2×3.5.

18.2×10^7 Add exponents.

1.82×10^8 Move decimal point and increase the exponent by 1.

Example:

$\dfrac{(4.1076 \times 10^3)}{2.8 \times 10^{-4}}$ Divide 4.1076 by 2.8.

1.467×10^7 Subtract $3 - (-4)$.

Radicals

RADICALS indicate an nth root of a number—for instance, $\sqrt[n]{a}$ is the nth root of a. Radical expressions are inverses of exponential expressions. As a result, $b = \sqrt[n]{a}$ and $b^n = a$ are equivalent. Based on this understanding, it is possible to estimate the value of b given a and n. Generally, the easiest way to find the exact (or nearly exact) value of b is to use a calculator, but the value can be estimated first by finding the nearest whole number and then, as necessary, increasing the precision of the estimate.

> **RADICALS:** indicate an nth root of a number

Example: Estimate the value of $\sqrt[3]{7}$.

First, express the problem in exponential form, and then determine a range for b.

$b = \sqrt[3]{7}$

$b^3 = 7$

$1^3 = 1,\ 2^3 = 8$

At this point, it is clear that the answer b must be between 1 and 2 (and it is clear that it is closer to 2 than to 1). Try $b = 1.9$. The result of b^3 can be found by hand (although it is slightly more difficult than dealing with whole numbers). The result is 6.859. Thus, it is apparent that b is between 1.9 and 2. Further refinement of this estimate is possible but time-consuming if done by hand.

Example: Estimate the value of $\sqrt{64}$.
$$b = \sqrt{64}$$
$$b^2 = 64$$
$$b = 8$$

This example follows the same procedure as that used previously, but the exact answer is a whole number and can be more easily determined than an irrational decimal value.

SKILL 1.9   Find powers and roots

Roots and powers are exact opposites. A power is the same as an exponent: You multiply the number by itself as many times as the power indicates.

Example:
$$4^3 = 64 \ (4 \times 4 \times 4)$$
$$7^2 = 49 \ (7 \times 7)$$
$$5^5 = 3125 \ (5 \times 5 \times 5 \times 5 \times 5)$$

Roots are asking you, "What number needs to be multiplied by itself the number of times indicated by the root in order to equal the number in question?"

Example:
The third root of $27 = 3$ $(3 \times 3 \times 3 = 27)$.
The fifth root of $3125 = 5$ $(5 \times 5 \times 5 \times 5 \times 5 = 3125)$.

Therefore, to find the fifth root of 64, ask, "What number multiplied by itself 5 times equals 64?" The answer is approximately 2.297. A calculator was required to find the answer. There is no simple way to calculate more complex roots by hand. You can see that whole powers—say, n—of any number—say, x—are easily calculated by hand: just multiply x by itself n times. Calculations are much more difficult when you want to find a fractional root or raise a number to a fractional power.

Given newly defined operations on a number system, determine whether the closure, commutative, associative, or distributive properties hold

The associative property holds for both addition and multiplication. You can multiply and add in any order:

$3(4 \times 2) = (3 \times 4)2$ and $3 + (4 + 2) = (3 + 4) + 2$

However, subtraction is not associative: $(8 - 3) - 2 \neq 8 - (3 - 2)$ and neither is division: $(16 \div 4) \div 2 \neq 16 \div (4 \div 2)$

The commutative property holds for addition and multiplication but not for subtraction and division: $a + b = b + a$.

$3 + 2 = 2 + 3$ and $3 \times 2 = 2 \times 3$, but $3 - 2 \neq 2 - 3$ and $3 \div 2 \neq 2 \div 3$

The distributive property holds for addition:

$a \times (b + c) = (a \times b) + (a \times c)$, and multiplication:
$(a + b) \times c = (a \times c) + (b \times c)$

$3(2 + 4) = 3 \times 2 + 3 \times 4$; however, $3 \div (2 - 1) \neq (3 \div 2) - (3 \div 1)$. The distributive property does not hold for division or subtraction.

The closure property says that if you add any two real numbers, you will get a real number. This property holds for addition, multiplication, and subtraction, but not division, since you cannot divide a real number by zero.

PROPERTIES OF REAL NUMBERS		
NAME	ADDITION	MULTIPLICATION
Commutativity	$a + b = b + a$	$ab = ba$
Associativity	$(a + b) + c = a + (b + c)$	$(ab)c = a(bc)$
Distributivity	$a(b + c) = ab + bc$	$(a + b)c = ac + bc$
Identity	$a + 0 = a = 0 + a$	$a \times 1 = a = 1 \times a$
Inverses	$a + (-a) = 0 = (-a) + a$	$a(a^{-1}) = 1$ if $a \neq 0$

The distributive property of exponents over addition does *not* hold:

$(a + b)^2 \neq a^2 + b^2$. If you consider that $10^2 = 100$, does it seem logical that $11^2 = 101$? No:
$11^2 = (10 + 1)^2 \neq 10^2 + 1^2 \neq 101$.

You know this intuitively when you are thinking of real numbers. However, the commutative property does hold for exponents: $(ab)^2 = (ba)^2$

Prime and Composite Numbers

PRIME NUMBERS: numbers that can be factored only into 1 and the number itself

PRIME NUMBERS are numbers that can be factored only into 1 and the number itself. When factoring into prime factors, all the factors must be numbers that cannot be factored again (without using 1). Initially numbers can be factored into any two factors. Check each resulting factor to see if it can be factored again. Continue factoring until all of the remaining factors are prime, and these are the prime factors. Regardless of what way the original number was factored, the final list of prime factors will always be the same. Prime numbers are whole numbers greater than 1 that have only two factors: 1 and the number itself. Examples of prime numbers are 2, 3, 5, 7, 11, 13, 17, or 19. Note that 2 is the only even prime number.

Example: Factor 30 into prime factors.

Factor 30 into any two factors:

    5×6 Now factor the 6.

    $5 \times 2 \times 3$ These are all prime factors.

Factor 30 into any two factors:

    3×10 Now factor the 10.

    $3 \times 2 \times 5$ These are the same prime factors even though the original factors were different.

Example: Factor 240 into prime factors.

Factor 240 into any two factors:

    24×10 Now factor both 24 and 10.

    $4 \times 6 \times 2 \times 5$ Now factor both 4 and 6.

    $2 \times 2 \times 2 \times 3 \times 2 \times 5$ These are prime factors.

This can also be written as $2^4 \times 3 \times 5$.

COMPOSITE NUMBERS: whole numbers that have more than two factors

COMPOSITE NUMBERS are whole numbers that have more than two factors. For example, 9 is composite because in addition to the factors 1 and 9, it has factor 3. 70 is also composite because in addition to the factors of 1 and 70, it has factors 2, 5, 7, 10, 14, and 35.

Remember that the number 1 is neither prime nor composite.

Even and Odd Numbers

Even numbers

Even numbers are numbers that are divisible by 2 with no remainder. They are divided into groups of two. For example, the number 6 can be divided into three groups of two.

Even numbers always end with an even digit such as 0, 2, 4, 6, or 8. In general, an even number is written as $n = 2K$, where K is an integer.

The set of even numbers is denoted by
$\{\text{Evens}\} = 2K = \{\ldots, -6, -4, -2, 0, 2, 4, 6, \ldots\}$

Counting even numbers by twos

We can start from 0 and count by twos in either of the following ways to get an even number:

- Add 2 to the previous number

- Count and skip every other number

Odd numbers

Odd numbers are the numbers that are not divisible by 2—that is, when an odd number is divided by 2, there is a remainder of 1.

Odd numbers always end with an odd digit such as 1, 3, 5, 7, and 9. The general form of an odd number is $n = 2K + 1$, where K is an integer.

The set of odd numbers can be written as
$\{\text{Odds}\} = 2K + 1 = \{\ldots, -9, -7, -5, -3, -1, 1, 3, 5, 7, 9, \ldots\}$

Counting odd numbers by twos

We can start from 1 and count by twos to get the odd number in either of the following ways:

- Add 2 to the previous number

- Count and skip every other number

Arithmetic operations on even and odd numbers

The following are the rules on how to add, subtract, multiply, and divide odd and even numbers. Even numbers can be written in the form $2n$ and odd numbers as $2n + 1$ where n is an integer. In the sections below, m and n stand for any integer.

Addition and Subtraction

Even ± Even = Even

$2n \pm 2m = 2(n \pm m)$

Example:

$12 + 6 = 18$

$6 - 4 = 2$

Odd ± Odd = Even

$2n + 1 \pm (2m + 1) = 2(n \pm m) + 1 \pm 1$

Example:

$9 + 7 = 16$

$13 - 9 = 4$

Even ± Odd = Odd

$2n \pm (2m + 1) = 2(n \pm m) \pm 1$

Example:

$14 + 9 = 23$

$28 - 3 = 25$

Multiplication

Even × Even = Even

$2n \times 2m = 4nm$

Example:

$4 \times 6 = 24$

Even × Odd = Even

$2n \times (2m + 1) = 2n(2m + 1)$

Example:

$2 \times 9 = 18$

Odd × Odd = Odd

$(2n + 1) \times (2m + 1) = 4mn + 2(n + m) + 1$

Example:

$5 \times 7 = 35$

Division

One cannot have general rules for division in all cases since division often produces fractions that are neither even nor odd. In cases where the quotient of a division is a whole number, the following rules apply:

Even ÷ Even = Even or Odd

(The quotient will be even if the dividend has more factors of 2 than the divisor.)

Example:
$$24 \div 6 = 4$$
$$24 \div 8 = 3$$

Even ÷ Odd = Even

(Since the even dividend has at least one factor of 2 and only odd factors will be cancelled by the divisor, at least one factor of 2 will remain.)

Example:
$$12 \div 3 = 4$$

Odd ÷ Odd = Odd

(Neither dividend nor divisor has any factors of 2.)

Example:
$$33 \div 11 = 3$$

Factors and Multiples

GCF

GCF stands for the GREATEST COMMON FACTOR, which is the largest number that is a factor of all the numbers given in a problem. The GCF can be no larger than the smallest number given in the problem. If no other number is a common factor, then the GCF will be the number 1. To find the GCF, list all of the possible factors of the smallest number given (include the number itself). Starting with the largest factor (which is the number itself), determine if it is also a factor of all the other given numbers. If so, that is the GCF. If that factor does not work, try the same method on the next smaller factor. Continue until a common factor is found. That is the GCF.

Example: Find the GCF of 12, 20, and 36.
The smallest number in the problem is 12. The factors of 12 are 1, 2, 3, 4, 6, and 12. The largest factor is 12, but it does not divide evenly into 20. Neither does 6, but 4 will divide into both 20 and 36 evenly.

Therefore, 4 is the GCF.

Example: Find the GCF of 14 and 15.
The factors of 14 are 1, 2, 7, and 14. The largest factor is 14, but it does not divide evenly into 15. Neither does 7 or 2. Therefore, the only factor common to both 14 and 15 is the number 1, which is the GCF.

> **GREATEST COMMON FACTOR:** the largest number that is a factor of all the numbers given in a problem

> *Note: There can be other common factors besides the GCF.*

LCM

> **LEAST COMMON MULTIPLE:** the smallest number into which all of the given numbers will divide

LCM is the abbreviation for LEAST COMMON MULTIPLE. The least common multiple of a group of numbers is the smallest number into which all of the given numbers will divide. The least common multiple will always be the largest of the given numbers or a multiple of the largest number.

Example: Find the LCM of 20, 30, and 40.

The largest number given is 40, but 30 will not divide evenly into 40. The next multiple of 40 is 80 (2 × 40), but 30 will not divide evenly into 80 either. The next multiple of 40 is 120. 120 is divisible by both 20 and 30, so 120 is the LCM (least common multiple).

Example: Find the LCM of 96, 16, and 24.

The largest number is 96. The number 96 is divisible by both 16 and 24, so 96 is the LCM.

Divisibility Tests

Test	Example
A number is divisible by 2 if that number is an even number (which means it ends in 0, 2, 4, 6, or 8)	1354 ends in 4, so it is divisible by 2. 240,685 ends in a 5, so it is not divisible by 2.
A number is divisible by 3 if the sum of its digits is evenly divisible by 3	The sum of the digits of 964 is 9 + 6 + 4 = 19. Since 19 is not divisible by 3, neither is 964. The digits of 86,514 are 8 + 6 + 5 + 1 + 4 = 24. Since 24 is divisible by 3, 86,514 is also divisible by 3.
A number is divisible by 4 if the number in its last two places is evenly divisible by 4	The number 113,336 ends with the number 36 in the last two places. Since 36 is divisible by 4, then 113,336 is also divisible by 4. The number 135,627 has the number 27 in its last two places. Since 27 is not evenly divisible by 4, then 135,627 is also not divisible by 4.
A number is divisible by 5 if the number ends in either a 5 or a 0	225 ends with a 5, so it is divisible by 5. The number 470 is also divisible by 5 because its last digit is a 0. 2,358 is not divisible by 5 because its last digit is an 8, not a 5 or a 0.
A number is divisible by 6 if the number is even and the sum of its digits is evenly divisible by 3	4,950 is an even number, and its digits add to 18 (4 + 9 + 5 + 0 = 18). Since the number is even and the sum of its digits is 18 (which is divisible by 3), then 4,950 is divisible by 6. 326 is an even number, but its digits add up to 11. Since 11 is not divisible by 3, then 326 is not divisible by 6. 135 is not an even number, so it cannot possibly be divided evenly by 6.

Table continued on next page

A number is divisible by 8 if the number in its last three digits is evenly divisible by 8	The number 113,336 ends with the three-digit number 336 in the last three places. Since 336 is divisible by 8, then 113,336 is also divisible by 8. The number 465,627 ends with the number 627 in the last three places. Since 627 is not evenly divisible by 8, then 465,627 is also not divisible by 8.
A number is divisible by 9 if the sum of its digits is evenly divisible by 9	The sum of the digits of 874 is $8 + 7 + 4 = 19$. Since 19 is not divisible by 9, neither is 874. The digits of 116,514 are $1 + 1 + 6 + 5 + 1 + 4 = 18$. Since 18 is divisible by 9, 116,514 is also divisible by 9.
A number is divisible by 10 if the number ends in the digit 0	305 ends with a 5, so it is not divisible by 10. The number 2,030,270 is divisible by 10 because its last digit is a 0. 42,978 is not divisible by 10 because its last digit is an 8, not a 0.

Why the divisibility tests work

All even numbers are divisible by 2 by definition.

A two-digit number (with T as the tens digit and U as the ones digit) has as its sum of the digits $T + U$. Suppose this sum of $T + U$ is divisible by 3. Then it equals three times some constant K. So, $T + U = 3K$. Solving this for U, $U = 3K - T$. The original two-digit number would be represented by $10T + U$. Substituting $3K - T$ in place of U, this two-digit number becomes $10T + U = 10T + (3K - T) = 9T + 3K$. This two-digit number is clearly divisible by 3, since each term is divisible by 3. Therefore, if the sum of the digits of a number is divisible by 3, the number itself is also divisible by 3.

Since 4 divides evenly into 100, 200, or 300, 4 will divide evenly into any number of hundreds. The only part of a number that determines if 4 will divide into it evenly is the number in the last 2 places.

Numbers that are divisible by 5 end in 5 or 0. This is clear if you look at the answers to the multiplication table for 5.

Answers to the multiplication table for 6 are all even numbers. Since 6 factors into 2 times 3, the divisibility rules for 2 and 3 must both work.

Any number of thousands is divisible by 8. Only the last three places of the number determine whether it is divisible by 8.

A two-digit number (with T as the tens digit and U as the ones digit) has as its sum of the digits $T + U$. Suppose this sum of $T + U$ is divisible by 9. Then it equals 9 times some constant K. So, $T + U = 9K$. Solving this for U, $U = 9K - T$. The original two-digit number would be represented by $10T + U$. Substituting $9K - T$ in place of U, this two-digit number becomes $10T + U = 10T + (9K - T) = 9T + 9K$. This two-digit number is clearly divisible by 9, since each term is divisible by 9. Therefore, if the sum of the digits of a number is divisible by 9, then the number itself is also divisible by 9.

Numbers divisible by 10 must be multiples of 10, which all end in a zero.

SKILL 1.12	Interpret and apply the concepts of ratio, proportion, and percent in appropriate situations

Consumer Applications

UNIT RATE: when purchasing an item; the price of that item divided by the number of units of measure

The UNIT RATE when purchasing an item is its price divided by the number of units of measure (pounds, ounces, etc.) in the item. The item with the lower unit rate has the lower price.

Example: Find the item with the best unit price:

$1.79 for 10 ounces
$1.89 for 12 ounces
$5.49 for 32 ounces

$\frac{1.79}{10} = 0.179$ per ounce $\frac{1.89}{12} = 0.1575$ per ounce $\frac{5.49}{32} = 0.172$ per ounce

$1.89 for 12 ounces is the best price.

A second way to find the better buy is to make a proportion with the price over the number of ounces (or whatever). Cross-multiply the proportion, writing the products above the numerator that is used. The better price will have the smaller product.

Example: Find the better buy: $8.19 for forty pounds or $4.89 for twenty-two pounds. Find the unit costs.

$$\frac{40}{8.19} = \frac{1}{x}$$
$$40x = 8.19$$
$$x = 0.20475$$

$$\frac{22}{4.89} = \frac{1}{x}$$
$$22x = 4.89$$
$$x = 0.222\overline{27}$$

Since $0.20475 < 0.222\overline{27}$, $8.19 is the lower price and a better buy.

To find the amount of sales tax on an item, change the percent of sales tax into an equivalent decimal number. Then multiply the decimal number times the price of the object to find the sales tax. The total cost of an item will be the price of the item plus the sales tax.

Example: A guitar costs $120.00 plus 7% sales tax. How much are the sales tax and the total cost?

7% = .07 as a decimal

(.07)(120) = $8.40 sales tax

$120.00 + $8.40 = $128.40 ← total price

An alternative method to find the total cost is to multiply the price times the factor 1.07 (price + sales tax):

$120 × 1.07 = $8.40

This gives you the total cost in fewer steps.

Example: A suit costs $450.00 plus $6\frac{1}{2}$% sales tax. How much are the sales tax and the total cost?

$6\frac{1}{2}$% = .065 as a decimal

(.065)(450) = $29.25 sales tax

$450.00 + $29.25 = $479.25 ← total price

An alternative method to find the total cost is to multiply the price times the factor 1.065 (price + sales tax):

$450 × 1.065 = $479.25

This gives you the total cost in fewer steps.

Ratios

A RATIO is a comparison of two numbers. If a class has 11 boys and 14 girls, the ratio of boys to girls could be written one of three ways:

11:14   or   11 to 14   or $\frac{11}{14}$

> **RATIO:** a comparison of two numbers

The ratio of girls to boys is

14:11   or   14 to 11   or $\frac{14}{11}$

Ratios can be reduced when possible. A ratio of 12 cats to 18 dogs would reduce to 2:3, 2 to 3, or $\frac{2}{3}$.

Note: Read ratio questions carefully. Given a group of 6 adults and 5 children, the ratio of children to the entire group would be 5:11.

> **PROPORTION:** an equation in which a fraction is set equal to another

Proportions

A PROPORTION is an equation in which a fraction is set equal to another. To solve the proportion, multiply each numerator times the other fraction's denominator. Set these two products equal to each other and solve the resulting equation. This is called CROSS-MULTIPLYING the proportion.

> **CROSS-MULTIPLYING:** multiplying one fraction's numerator times another fraction's denominator

Example: $\frac{4}{15} = \frac{x}{60}$ *is a proportion.*

To solve this, cross-multiply:

$$(4)(60) = (15)(x)$$
$$240 = 15x$$
$$16 = x$$

Example: $\frac{x+3}{3x+4} = \frac{2}{5}$ *is a proportion.*

To solve, cross-multiply:

$$5(x + 3) = 2(3x + 4)$$
$$5x + 15 = 6x + 8$$
$$7 = x$$

Example: $\frac{x+2}{8} = \frac{2}{x-4}$ *is another proportion.*

To solve, cross-multiply:

$$(x + 2)(x - 4) = 8(2)$$
$$x^2 - 2x - 8 = 16$$
$$x^2 - 2x - 24 = 0$$
$$(x - 6)(x + 4) = 0$$
$$x = 6 \text{ or } x = -4$$

Both answers work.

SKILL 1.13 Recognize the reasonableness of results within the context of a given problem; using estimation, test the reasonableness of results

Estimation and approximation may be used to check the reasonableness of answers. This is particularly important when calculators are used. Students need to be able to verify that the answer they are getting by punching in numbers is in the correct range and makes sense in the given context. Estimation requires good mental math skills. There are several different ways of estimating.

A simple check for reasonableness is to ask whether the answer expected is *more or less* than a given number. It is astonishing how many errors of computation can be avoided using this method. For instance, when converting 20 Km to meters, ask whether you are expecting a number greater than or less than 20. That will tell you whether to multiply or divide by 1,000 (a common point of confusion in conversion problems).

Estimation Methods

The most common estimation strategies taught in schools involve replacing numbers with numbers that are easier to work with. These methods include rounding off, front-end digit estimation, and compensation. While rounding off is done to a specific place value (e.g., nearest ten or hundred), FRONT-END ESTIMATION involves rounding off or truncating to whatever place value the first digit in a number represents. The following example uses front-end estimation.

Example: Estimate the answer.

$$\frac{58 \times 810}{1{,}989}$$

58 becomes 60, 810 becomes 800, and 1,989 becomes 2,000.

$$\frac{60 \times 800}{2{,}000} = 24$$

COMPENSATION involves replacing different numbers in different ways so one change can more or less compensate for the other.

Example: 32 + 53 = 30 + 55 = 85

Here, both numbers are replaced in a way that minimizes the change; one number is increased and the other is decreased.

Another estimation strategy is to estimate a range for the correct answer.
Example: 458 + 873 > 400 + 800 and 458 + 873 < 500 + 900.
One can estimate that the sum of 458 and 873 lies in the range 1,200 to 1,400.

Converting to an equivalent fraction, decimal, or percentage can often be helpful.
Example: To calculate 25% of 520, first understand that 25% = $\frac{1}{4}$ and simply divide 520 by 4 to get 130.

CLUSTERING is a useful strategy when dealing with a set of numbers.

Similar numbers can be grouped together to simplify computation.
Example: 1210 + 655 + 1178 + 683 + 628 + 1223 + 599 = 600 + 600 + 600 + 600 + 1200 + 1200 + 1200.

Grouping together COMPATIBLE NUMBERS is a variant of clustering. Here, instead of similar numbers, numbers that together produce easy-to-compute numbers are grouped together.
Example: 5 + 17 + 25 + 23 + 40 = (5 + 25) + (17 + 23) + 40.

FRONT-END ESTIMATION: involves rounding off or truncating to whatever place value the first digit in a number represents

COMPENSATION: involves replacing different numbers in different ways so one change can more or less compensate for the other

CLUSTERING: grouping numbers together to simplify computations

COMPATIBLE NUMBERS: grouping together numbers that produce easy-to-compute numbers

Applications of Estimation

Often a problem does not require exact computation. An estimate may sometimes be all that is needed to solve a word problem, as in the following example. Therefore, assessing what level of precision is needed in a particular situation is an important skill that must be taught to all students.

Example: Janet goes into a store to purchase a CD that is on sale for $13.95. While shopping, she sees two pairs of shoes, one pair priced $19.95 and one pair priced $14.50. She only has $50. Can she purchase the CD and both pairs of shoes? (Assume there is no sales tax.)

Solve by rounding:

$19.95 → $20.00

$14.50 → $15.00

$13.95 → $14.00

$49.00   Yes, she can purchase the CD and the shoes.

Elapsed time problems are usually one of two types: the elapsed time between two times given in hours, minutes, and seconds, and between two times given in months and years.

For any time of day past noon, change it into military time by adding 12 hours. For instance, 1:15 p.m. would be 13:15. Remember that when you borrow a minute or an hour in a subtraction problem, you have borrowed 60 more seconds or minutes.

Example: Find the time from 11:34:22 a.m. until 3:28:40 p.m.

First, change 3:28:40 p.m. to 15:28:40 p.m.

Now subtract   − 11:34:22 a.m.

                                            :18

Borrow an hour and add 60 more minutes. Subtract.

14:88:40 p.m.

− 11:34:22 a.m.

3:54:18 ↔ 3 hours, 54 minutes, 18 seconds

Example: Jermaine lived in Arizona from September 1991 until March 1995. How long did he live there?

		Year	Month
March 1995	=	95	03
September 1991	= −	91	09

Borrow a year, change it into 12 more months, and *subtract*.

		Year	Month
March 1995	=	94	15
September 1991	=	− 91	09
		3 years	6 months

Example: A race took the winner 1 hr. 58 min. 12 sec. on the first half of the race and 2 hr. 9 min. 57 sec. on the second half of the race. How much time did the entire race take?

1 hr 58 min 12 sec	
+ 2 hr 9 min 57 sec	Add these numbers.
3 hr 67 min 69 sec	
+ 1 min − 60 sec	Change 60 sec to 1 min.
3 hr 68 min 9 sec	
+ 1 hr − 60 min	Change 60 min to 1 hr.
4 hr 8 min 9 sec	Final answer.

> ### SKILL 1.14 Work with algebraic expressions, formulas, and equations; add, subtract, and multiply polynomials; divide polynomials

For coverage of working with algebraic expressions, formulas, and equations, see Skill 1.15

Finding the LCD of Two Expressions

In order to add or subtract rational expressions, they must have a common denominator. If they don't have a common denominator, then factor the denominators to determine what factors are missing from each denominator to make the LCD. Multiply both the numerator and denominator by the missing factor(s). Once the fractions have a common denominator, add or subtract their numerators, but keep the common denominator the same. Factor the numerator if possible, and reduce if there are any factors that can be canceled.

Example: Find the least common denominator for $6a^3b^2$ and $4ab^3$.

These factor into $2 \times 3 \times a^3 \times b^2$ and $2 \times 2 \times a \times b^3$.

The first expression needs to be multiplied by another 2 and b.

The other expression needs to be multiplied by 3 and a^2.

Then both expressions would be $2 \times 2 \times 3 \times a^3 \times b^3 = 12a^3b^3 = $ LCD

Adding Polynomials

We find the sum of two polynomials by grouping the like powers, retaining their signs and adding the coefficients of like powers. There are different ways to add polynomials.

Using the horizontal method to add like terms

Remove the parentheses, and identify like terms. Group the like terms together. Add the like terms.

Example: Add $(2x^2 - 4)$ and $(x^2 + 3x - 3)$.

$$(2x^2 - 4) + (x^2 + 3x - 3) = 2x^2 + x^2 + 3x - 4 - 3$$
$$= 3x^2 + 3x - 7$$

Example: Add $(5x - 1) + (10x^2 + 7x)$.

$$(5x - 1) + (10x^2 + 7x) = 10x^2 + 5x + 7x - 1$$
$$= 10x^2 + 12x - 1$$

Example: Add $(20x^2 + 2) + (15x^2 - 8) + (3x^2 - 4)$.

$$(20x^2 + 2) + (15x^2 - 8) + (3x^2 - 4) = 20x^2 + 15x^2 + 3x^2 + 2 - 8 - 4$$
$$= 38x^2 - 10$$

Using the vertical method to add like terms

Arrange the like terms so they are lined up under one another in vertical columns. Add the like terms in each column, following the rules for adding the signed numbers.

Example: Add $(2x^2 - 4)$ and $(x^2 + 3x - 3)$.

$$\begin{array}{r} 2x^2 + 0x - 4 \\ \underline{x^2 + 3x - 3} \\ 3x^2 + 3x - 7 \end{array}$$

Example: Add $(x^4 - 3x^3 + 2x + 6)$ and $(x^3 + 2)$.

$$\begin{array}{r} x^4 - 3x^3 + 2x + 6 \\ \underline{0x^4 + x^3 + 0x + 2} \\ x^4 - 2x^3 + 2x + 8 \end{array}$$

Example: Add $(3y^7 - 2y^2 + 3)$ and $(y^6 - 3y^4 + y^2 + y)$.

$$3y^7 - 0y^6 + 0y^5 + 0y^4 + 0y^3 - 2y^2 + 0y + 3$$
$$\underline{0y^7 + 1y^6 + 0y^5 - 3y^4 + 0y^3 + 1y^2 + 1y + 0}$$
$$3y^7 + y^6 + 0y^5 - 3y^4 + 0y^3 - y^2 + y + 3$$

The answer is $3y^7 + y^6 - 3y^4 - y^2 + y + 3$.

Subtracting Polynomials

Subtract like terms by changing the signs of the terms being subtracted and following the rules for addition of polynomials.

Example: Subtract $(y^4 + 4y^2 + 5y - 4)$ from $(3y^5 + y^4 - 3y^3 - 2y + 1)$.
First, we change the sign of each term of the polynomial being subtracted. So the polynomial $y^4 + 4y^2 + 5y - 4$ becomes $-y^4 - 4y^2 - 5y + 4$.

$$(3y^5 + y^4 - 3y^3 - 2y + 1) - (y^4 + 4y^2 + 5y - 4)$$
$$= (3y^5 + y^4 - 3y^3 - 2y + 1) + (-y^4 - 4y^2 - 5y + 4)$$
$$= 3y^5 + 0y^4 - 3y^3 - 7y + 5$$
$$= 3y^5 - 3y^3 - 7y + 5$$

Example: Subtract $(p^3 + 4p^2 - 5p - 6)$ from $(7p^3 - 3p + 5)$.
$$(7p^3 - 3p + 5) - (p^3 + 4p^2 - 5p - 6) = 7p^3 - p^3 - 4p^2 - 3p + 5p + 5 + 6$$
$$= 6p^3 - 4p^2 + 2p + 11$$

Using the vertical method to subtract polynomials

We change the sign of each term of the polynomial being subtracted and add with the other polynomial.

Example: Subtract $(t^3 - 5t + 1)$ from $(t^4 + 3t^3 - 5t - 1)$.

$$t^4 + 3t^3 - 5t - 1$$
$$\underline{0t^4 - t^3 + 5t - 1}$$
$$t^4 + 2t^3 + 0t - 2$$

The answer is $t^4 + 2t^3 - 2$.

Example: Subtract ($p^3 + 4p^2 - 5p - 6$) from ($7p^3 - 3p + 5$).

After changing the sign of each term of $p^3 + 4p^2 - 5p - 6$, we have to add with $7p^3 + 0p^2 - 3p + 5$.

$$
\begin{array}{r}
7p^3 + 0p^2 - 3p + 5 \\
\underline{-p^3 - 4p^2 + 5p + 6} \\
6p^3 - 4p^2 + 2p + 11
\end{array}
$$

Multiplying Polynomials

All of the terms of the second polynomial are to be multiplied by each term of the first polynomial to get the product.

Example: Simplify ($5x$)($2x^3$).

$$
\begin{aligned}
(5x)(2x^3) &= (5 \times 2)(x \times x^3) \\
&= 10x^4
\end{aligned}
$$

Example: Simplify ($-3x$)($4x^2 - x + 10$).

$$
\begin{aligned}
(-3x)(4x^2 - x + 10) &= (-3x \times 4x^2) - (-3x \times x) + (-3x \times 10) \\
&= -12x^3 - -3x^2 + -30x \\
&= -12x^3 + 3x^2 - 30x
\end{aligned}
$$

Example: Simplify ($y + 2$)($y - 3$).

$$
\begin{aligned}
(y + 2)(y - 3) &= (y \times y) - (y \times 3) + (2 \times y) - (2 \times 3) \\
&= y^2 - 3y + 2y - 6 \\
&= y^2 - y - 6
\end{aligned}
$$

Example: Find the product of ($2p + 6$)($2p - 6$).

$$
\begin{aligned}
(2p + 6)(2p - 6) &= (2p \times 2p) - (2p \times 6) + (6 \times 2p) - (6 \times 6) \\
&= 4p^2 - 12p^2 + 12p - 36 \\
&= 4p^2 - 36
\end{aligned}
$$

Dividing Polynomials

One polynomial can be divided by another using long division that is similar to long division performed with numbers, i.e. start working with the front part of the dividend and bring down one term at a time. The examples given below will clarify the procedure.

Example: Divide $2x^2 + 5x - 1$ by $2x - 1$.

$$\begin{array}{r} x \\ 2x-1\overline{)2x^2+5x-1} \\ \underline{-(2x^2-x)} \\ 6x-1 \end{array}$$

First, note that $2x$ times x will give $2x^2$

Bring down the -1

$$\begin{array}{r} x+3 \\ 2x-1\overline{)2x^2+5x-1} \\ \underline{-(2x^2-x)} \\ 6x-1 \\ \underline{-(6x-3)} \\ 2 \end{array}$$

Note that $2x$ times 3 gives $6x$

The quotient is $x + 3$ and the remainder is 2.

Example: Divide $2t^3 - 9t^2 + 11t - 3$ by $2t - 3$.

$$\begin{array}{r} t^2-3t+1 \\ 2t-3\overline{)2t^3-9t^2+11t-3} \\ \underline{-(2t^3-3t^2)} \\ -6t^2+11t \\ \underline{-(-6t^2+9t)} \\ 2t-3 \\ \underline{-(2t-3)} \\ 0 \end{array}$$

The quotient is $t^2 - 3t + 1$ and the remainder is zero.

If the divisor is of the form $x - a$, synthetic division may be used to divide polynomials. For details about synthetic division see Skill 1.20.

SKILL 1.15 · Add, subtract, multiply, and divide algebraic fractions

Finding the LCD

Example: Find the LCD for $x^2 - 4$, $x^2 + 5x + 6$, and $x^2 + x - 6$.

$x^2 - 4$ factors into $(x - 2)(x + 2)$
$x^2 + 5x + 6$ factors into $(x + 3)(x + 2)$
$x^2 + x - 6$ factors into $(x + 3)(x + 2)$

The LCD is $(x + 3)(x + 2)(x - 2)$.

Adding Algebraic Fractions

To add algebraic fractions, first express the fractions in terms of their lowest common denominator (LCD), and then add their numerators to obtain the numerator that is divided by the LCD to arrive at the answer.

Example: Add $\left(\frac{3}{a}\right)$ and $\left(\frac{5a}{3}\right)$.

$$\left(\tfrac{3a}{2}\right) + \left(\tfrac{5a}{3}\right) = \left(\tfrac{3a \times 3}{6}\right) + \left(\tfrac{5a \times 2}{6}\right)$$
$$= \tfrac{9a}{6} + \tfrac{10a}{6}$$
$$= \tfrac{19a}{6} = 3\tfrac{1}{6}a$$

Example: Find the sum of $\left(\frac{5}{x+y}\right)$ and $\left(\frac{4}{x-y}\right)$.

$$\left(\tfrac{5}{x+y}\right) + \left(\tfrac{5}{x-y}\right) = \left[\tfrac{5(x-y)}{(x+y)(x-y)}\right] + \left[\tfrac{4(x+y)}{(x+y)(x-y)}\right]$$
$$= \tfrac{5(x-y) + 4(x+y)}{(x+y)(x-y)}$$
$$= \tfrac{5x - 5y + 4x + 4y}{x^2 - y^2}$$
$$= \tfrac{9x - y}{x^2 - y^2}$$

Example: Find the value of $\left(\frac{7x}{8}\right) + 4$.

$$\left(\tfrac{7x}{8}\right) + 4 = \left(\tfrac{7x \times 1}{8}\right) + \left(\tfrac{4 \times 8}{8}\right)$$
$$= \tfrac{7x + 32}{8}$$

Example:

$$\tfrac{5}{6a^3b^2} + \tfrac{1}{4ab^3} = \tfrac{5(2b)}{6a^3b^2(2b)} + \tfrac{1(3a^2)}{4ab^3(3a^2)} = \tfrac{10a}{12a^3b^3} + \tfrac{3b^2}{12a^3b^3} = \tfrac{10a + 3b^2}{12a^3b^3}$$

This will not reduce, since all three terms are not divisible by anything.

Subtracting Algebraic Fractions

To subtract algebraic fractions, first express the fractions in terms of their LCD and then subtract their numerators to obtain the numerator that is divided by the LCD to arrive at the answer.

Example:

$$\tfrac{2}{x^2 - 4} - \tfrac{3}{x^2 + 5x + 6} + \tfrac{7}{x^2 + x - 6} =$$
$$\tfrac{2}{(x-2)(x+2)} - \tfrac{3}{(x+3)(x+2)} + \tfrac{7}{(x+3)(x-2)} =$$
$$\tfrac{2(x+3)}{(x-2)(x+2)(x+3)} - \tfrac{3(x-2)}{(x+3)(x+2)(x-2)} + \tfrac{7(x+2)}{(x+3)(x-2)(x+2)} =$$
$$\tfrac{2x+6}{(x-2)(x+2)(x+3)} - \tfrac{3x-6}{(x+3)(x+2)(x-2)} + \tfrac{7x+14}{(x+3)(x-2)(x+2)} =$$
$$\tfrac{2x + 6 - (3x - 6) + 7x + 14}{(x+3)(x-2)(x+2)} = \tfrac{6x + 26}{(x+3)(x-2)(x+2)}$$

This will not reduce.

Example: Add $(\frac{3a}{2})$ and $(\frac{5a}{3})$.

$$(\tfrac{3a}{2}) + (\tfrac{5a}{3}) = (\tfrac{3a \times 3}{6}) + (\tfrac{5a \times 2}{6})$$
$$= \tfrac{9a}{6} + \tfrac{10a}{6}$$
$$= \tfrac{19a}{6} = 3\tfrac{1}{6}a$$

Example: $(\frac{5}{x+y})(\frac{4}{x-y})$.

$$\frac{5}{(x+y)} + \frac{4}{(x-y)} = [\frac{5(x-y)}{(x+y)(x-y)}] + [\frac{4(x+y)}{(x+y)(x-y)}]$$
$$= \frac{5(x-y) + 4(x-y)}{(x+y)(x-y)}$$
$$= \frac{5x - 5y + 4x + 4y)}{x^2 - y^2}$$
$$= \frac{9x - y}{x^2 - y^2}$$

Multiplying Algebraic Fractions

Factorize the numerators and denominators. Then cancel the factors that are common to the numerator and denominator before applying multiplication to obtain the answer.

Example: Find the product of $(\frac{a}{3})$ and $(\frac{b}{4})$.

$$(\tfrac{a}{3}) \times (\tfrac{b}{4}) = \tfrac{a \times b}{3 \times 4} = \tfrac{ab}{12}$$

Example: Simplify $(\frac{5x^2}{2x^2 - 2x})(\frac{x^2 - 1}{x^2 + x})$.

$(\frac{5x^2}{2x^2 - 2x})(\frac{x^2 - 1}{x^2 + x}) = [\frac{5x^2}{2x(x-1)}][\frac{(x+1)(x-1)}{x(x+1)}] = \frac{5}{2}$, since all of the common factors from the numerator and the denominator are canceled.

Example: Multiply $(\frac{4x^2}{x-2})(\frac{x^2-4}{12})$.

$$\left(\frac{4x^2}{x-2}\right)\left(\frac{x^2-4}{12}\right) = \left(\frac{4x^2}{x-2}\right)\left[\frac{(x-2)(x+2)}{(4 \times 3)}\right] = \frac{x^2(x+2)}{3}$$

Dividing Algebraic Fractions

To divide the algebraic fractions, invert the second fraction and multiply it by the first fraction. In other words, you must multiply both fractions after taking the reciprocal of the divisor. Then factorize the numerator and denominator and cancel all the common factors.

Example: Simplify $(\frac{x-3}{4}) \div (\frac{x+3}{8})$.

$$(\tfrac{x-3}{4}) \div (\tfrac{x+3}{8}) = (\tfrac{x-3}{4})(\tfrac{8}{x+3})$$
$$= \frac{2(x-3)}{x+3}$$

Example: Simplify $\left(\frac{m^2 + 2m}{3m^3}\right) \div \left(\frac{5m + 10}{6}\right).$

$$\left(\frac{m^2 + 2m}{3m^3}\right) \div \left(\frac{5m + 10}{6}\right) = \left[\frac{m(m + 2)}{3m^3}\right] \div \left[\frac{5(m + 2)}{6}\right]$$

$$= \left[\frac{m(m + 2)}{3m^3}\right]\left[\frac{6}{5(m + 2)}\right]$$

$$= \frac{2}{5m^2}$$

Example: Find $\left(\frac{12y + 24}{36}\right) \div \left(\frac{y^2 + 3y + 2}{2}\right).$

$$\left(\frac{12y + 24}{36}\right) \div \left(\frac{y^2 + 3y + 2}{2}\right) = \left[\frac{12(y + 2)}{12 \times 3}\right] \div \left[\frac{(y + 2)(y + 1)}{2}\right]$$

$$= \left(\frac{y + 2}{3}\right)\left[\frac{2}{(y + 2)(y + 1)}\right]$$

$$= \frac{2}{3(y + 1)}$$

Applications

Some problems can be solved using equations with rational expressions. First, write the equation. To solve it, multiply each term by the LCD of all of the fractions. This will cancel out all of the denominators and give an equivalent algebraic equation that can be solved.

Example: The denominator of a fraction is two less than three times the numerator. If 3 is added to both the numerator and the denominator, the new fraction equals $\frac{1}{2}$.

original fraction: $\frac{x}{3x - 2}$ revised fraction: $\frac{x + 3}{3x + 1}$

$$\frac{x + 3}{3x + 1} = \frac{1}{2}$$ $$2x + 6 = 3x + 1$$

$$x = 5$$

So the original fraction is $\frac{5}{13}$.

Example: Elly Mae can feed the animals in 15 minutes. Jethro can feed them in 10 minutes. How long will it take them if they work together?

If Elly Mae can feed the animals in 15 minutes, then she could feed $\frac{1}{15}$ of them in 1 minute, $\frac{2}{15}$ of them in 2 minutes, and $\frac{x}{15}$ of them in x minutes. In the same fashion, Jethro could feed $\frac{x}{10}$ of them in x minutes. Together they complete 1 job. The equation is:

$$\frac{x}{15} + \frac{x}{10} = 1$$

Multiply each term by the LCD (least common denominator) of 30:

$$2x + 3x = 30$$

$$x = 6 \text{ minutes}$$

Example: A salesman drove 480 miles from Pittsburgh to Hartford. The next day he returned the same distance to Pittsburgh in half an hour less time than his original trip took, because he increased his average speed by 4 mph. Find his original speed.

Since distance = rate × time, then time = $\frac{\text{distance}}{\text{rate}}$

original time − 1/2 hour = shorter return time

$\frac{480}{x} - \frac{1}{2} = \frac{480}{x+4}$

Multiplying by the LCD of $2x(x + 4)$, the equation becomes:

$480[2(x + 4)] - 1[x(x + 4)] = 480(2x)$

$960x + 3840 - x^2 - 4x = 960x$

$x^2 + 4x - 3840 = 0$

$(x + 64)(x - 60) = 0$

$x = 60$

60 mph is the original speed, 64 mph is the faster return speed.

Practice Set Problems

1. Working together, Larry, Moe, and Curly can paint an elephant in 3 minutes. Working alone, it would take Larry 10 minutes or Moe 6 minutes to paint the elephant. How long would it take Curly to paint the elephant if he worked alone?

2. The denominator of a fraction is 5 more than twice the numerator. If the numerator is doubled, and the denominator is increased by 5, the new fraction is equal to $\frac{1}{2}$. Find the original fraction.

3. A trip from Augusta, Maine, to Galveston, Texas, is 2,108 miles. If one car drove 6 mph faster than a truck and got to Galveston 3 hours before the truck, find the speeds of the car and the truck.

Answer Key

1. It takes Curly 15 minutes to paint the elephant alone.

2. The original fraction is $\frac{5}{15}$.

3. The car was traveling at 68 mph, and the truck was traveling at 62 mph.

Solving for a Variable

To solve an algebraic formula for some variable R, use the following steps:

1. Eliminate any parentheses using the distributive property

2. Multiply every term by the LCD of any fractions to write an equivalent equation without any fractions

3. Move all of the terms containing the variable R to one side of the equation. Move all of the terms without the variable to the opposite side of the equation

4. If there are two or more terms containing the variable R, factor only R out of each of those terms as a common factor

5. Divide both sides of the equation by the number or expression being multiplied times the variable R

6. Reduce the fractions if possible

7. Remember that there are restrictions on values allowed for variables because the denominator cannot equal zero

Examples:

Solve $A = p + prt$ for t.

$$A - p = prt$$
$$\frac{A - p}{pr} = \frac{prt}{pr}$$
$$\frac{A - p}{pr} = t$$

Solve $A = p + prt$ for p.

$$A = p(1 + rt)$$
$$\frac{A}{1 + rt} = \frac{p(1 + rt)}{1 + rt}$$
$$\frac{A}{1 + rt} = p$$

Solve $A = \frac{1}{2} h(b_1 + b_2)$ for b_2.

$A = \frac{1}{2} hb_1 + \frac{1}{2} hb_2$ ← step a

$2A = hb_1 + hb_2$ ← step b

$2A - hb_1 = hb_2$ ← step c

$\dfrac{2A - hb_1}{h} = \dfrac{hb_2}{h}$ ← step d

$\dfrac{2A - hb_1}{h} = b_2$ ← will not reduce

COMPLEX NUMBERS are of the form $a + bi$, where a and b are real numbers and $i = \sqrt{-1}$. When i appears in an answer, it is acceptable unless it is in a denominator. When i^2 appears in a problem, it is always replaced by -1. Remember: $i^2 = -1$.

> **COMPLEX NUMBERS:** numbers of the form $a + bi$, where a and b are real numbers and $i = \sqrt{-1}$

Adding Complex Numbers

To add complex numbers, add the real parts together and add the imaginary parts together.

Example: Add (3 + 4i) and (5 − 2i).

$$(3 + 4i) + (5 - 2i) = (3 + 5) + (4i - 2i)$$
$$= 8 + 2i$$

Example: Find the sum of (3 + 6i), (-5 + i), (9 − 4i), and (5i).

$$(3 + 6i) + (-5 + i) + (9 - 4i) + (5i) = (3 - 5 + 9) + (6i + i - 4i + 5i)$$
$$= 7 + 8i$$

Example: Simplify (-1 − i) + (4 − 2i) + (3i) + (10) + (7 + 6i).

$$(-1 - i) + (4 - 2i) + (3i) + (10) + (7 + 6i) =$$
$$(-1 + 4 + 0 + 10 + 7) + (-i + -2i + 3i + 6i) =$$
$$20 + 6i$$

Subtracting Complex Numbers

To subtract a complex number, change the sign of the real and imaginary parts and add.

Example 1: Simplify (2 + 3i) − (1 + 3i).

$$(2 + 3i) - (1 + 3i) = (2 + 3i) + (-1 - 3i)$$
$$= (2 - 1) + (3i - 3i)$$
$$= 1$$

Example 2: Subtract (5 − 3i) from (3 + 2i).

$$(3 + 2i) - (5 - 3i) = (3 + 2i) + (-5 + 3i)$$
$$= (3 - 5) + (2i + 3i)$$
$$= -2 + 5i$$

Example 3: Subtract (7i) + (2 − 2i) − (4 + 3i) − (12 − 3i).
$$= (2 − 4 − 12) + (7i − 2i − 3i + 3i)$$
$$= \text{-}14 + 5i$$

Multiplying Complex Numbers

Multiplication of complex numbers is just like multiplication of binomials. Note that $i^2 = \text{-}1$.

To multiply two complex numbers, FOIL the two numbers together. Replace i^2 with -1 and finish combining like terms. Answers should have the form $a + b\,i$.

Example: Multiply (8 + 3i)(6 − 2i). FOIL this.

$48 − 16i + 18i − 6i^2$	Let $i^2 = \text{-}1$
$48 − 16i + 18i − 6(\text{-}1)$	
$48 − 16i + 18i + 6$	
$54 + 2i$	This is the answer.

Example: Simplify (5 + 8i)² ← Write this out twice.

$(5 + 8i)(5 + 8i)$	FOIL this.
$25 + 40i + 40i + 64i^2$	Let $i^2 = \text{-}1$
$25 + 40i + 40i + 64(\text{-}1)$	
$25 + 40i + 40i − 64$	
$\text{-}39 + 80i$	This is the answer.

Example: Simplify (2 + i) (3 + 2i).
$$(2 + i)(3 + 2i) = 2 \times 3 + 2 \times 2i + i \times 3 + i \times 2i$$
$$= 6 + 4i + 3i − 2$$
$$= 4 + 7i$$

Example: Find the product of (1 − i), (3 + 4i) and (2 − 3i).
$$(1 − i)(3 + 4i)(2 − 3i) = (1 \times 3 + 1 \times 4i − i \times 3 − i \times 4i)(2 − 3i)$$
$$(3 + 4i − 3i + 4)(2 − 3i)$$
$$= (7 + i)(2 − 3i)$$
$$= 7 \times 2 − 7 \times 3i + 2i − i \times 3i$$
$$= 14 − 21i + 2i + 3$$
$$= 17 − 19i$$

Dividing Complex Numbers

When dividing two complex numbers, you must eliminate the complex number in the denominator. If the complex number in the denominator is of the form $b\,i$, multiply both the numerator and the denominator by i. Remember to replace i^2 with -1 and then continue simplifying the fraction.

Example: $\dfrac{2 + 3i}{5i}$
Mulitply this by $\dfrac{i}{i}$

$$\frac{2 + 3i}{5i} \times \frac{i}{i} = \frac{(2 + 3i)i}{5i \times i} = \frac{2i + 3i^2}{5i^2} = \frac{2i + 3(-1)}{-5} = \frac{-3 + 2i}{-5} = \frac{3 - 2i}{5}$$

If the complex number in the denominator is of the form $a + b\,i$, multiply both the numerator and the denominator by the conjugate of the denominator. The CONJUGATE of the denominator of a complex number is the same two terms with the opposite sign between the two terms (the real term does not change signs). The conjugate of $2 - 3i$ is $2 + 3i$. The conjugate of $-6 + 11i$ is $-6 - 11i$. Multiply the factors on the top and bottom of the fraction. Remember to replace i^2 with -1, combine like terms, and then continue simplifying the fraction.

> **CONJUGATE:** the conjugate of a complex number is the same two terms with the opposite sign between the two terms (the real term does not change signs)

Example: $\dfrac{4 + 7i}{6 - 5i}$
Multiply by $\dfrac{6 + 5i}{6 + 5i}$, the conjugate of the denominator.

$$\frac{(4 + 7i)}{(6 - 5i)} \times \frac{(6 + 5i)}{(6 + 5i)} = \frac{24 + 20i + 42i + 35i^2}{36 + 30i - 30i - 25i^2} = \frac{24 + 62i + 35(-1)}{36 - 25(-1)} = \frac{-11 + 62i}{61}$$

Example: $\dfrac{24}{-3 - 5i}$
Multiply by $\dfrac{-3 + 5i}{-3 + 5i}$, the conjugate of the denominator.

$$\frac{24}{(-3 - 5i)} \times \frac{-3 + 5i}{-3 + 5i} = \frac{-72 + 120i}{9 - 25i^2} = \frac{-72 + 120i}{9 + 25} = \frac{-72 + 120i}{34} = \frac{-36 + 60i}{17}$$

Divided everything by 2.

> SKILL **Perform standard operations involving radicals and exponents,**
> 1.17 **including fractional and negative exponents**

The Laws of Radicals	Examples
1. $(\sqrt[n]{a})^n = a$	$(\sqrt[3]{6})^3 = 6$
2. $\sqrt[n]{ab} = \sqrt[n]{a}\,\sqrt[n]{b}$	$\sqrt[3]{10} = \sqrt[3]{2 \times 5} = \sqrt[3]{2} \times \sqrt[3]{5}$

Table continued on next page

3. $\sqrt[n]{\dfrac{a}{b}} = \dfrac{\sqrt[n]{a}}{\sqrt[n]{b}}$	$\sqrt{\dfrac{3}{7}} = \dfrac{\sqrt{3}}{\sqrt{7}}$
4. $\sqrt[n]{a^m} = (\sqrt[n]{a})^m$	$(2\sqrt[3]{4})(3\sqrt[3]{16}) = (2 \times 3)(\sqrt[3]{4} \times \sqrt[3]{16})$ $= 6\sqrt[3]{4 \times 16} = 6\sqrt[3]{4^3} = 6 \times 4 = 24$
5. $\sqrt[m]{\sqrt[n]{a}} = \sqrt[mn]{a}$	$\sqrt[3]{\sqrt{5}} = \sqrt[3 \times 2]{5} = \sqrt[6]{5}$

Adding and Subtracting Radicals

Addition or subtraction of two or more radicals is achieved by reducing each radical to the simplest form and then combining terms with similar radicals.

Example: Simplify $\sqrt{32} - \sqrt{50} + \sqrt{18}$.
$$\sqrt{32} - \sqrt{50} + \sqrt{18} = 4\sqrt{2} - 5\sqrt{2} + 3\sqrt{2}$$
$$= 2\sqrt{2}$$

Example: Simplify $\sqrt[3]{432} + \sqrt[4]{625} - \sqrt{128}$.
$$\sqrt[3]{432} + \sqrt[4]{625} - \sqrt{128} = \sqrt[3]{2^4 \times 3^3} + \sqrt[4]{5^4} - \sqrt{2^7}$$
$$= 6\sqrt[3]{2} + 5 - 8\sqrt{2}$$

Multiplying and Dividing Radicals

Multiplication and division of radicals are performed using the laws of radicals.

Example: Find $\dfrac{\sqrt[3]{4}}{\sqrt{2}}$.
$$\frac{\sqrt[3]{4}}{\sqrt{2}} = \frac{\sqrt[3]{2^2}}{\sqrt{2}}$$
$$= \frac{2^{\frac{2}{3}}}{2^{\frac{1}{2}}}$$
$$= 2^{\frac{2}{3} - \frac{1}{2}}$$
$$= 2^{\frac{1}{6}}$$

Example: Simplify

$$\frac{\sqrt[4]{225}}{\sqrt[3]{81}} = \frac{\sqrt[4]{5^2 \times 3^2}}{\sqrt[3]{3^4}} = \frac{5^{\frac{2}{4}} \times 3^{\frac{2}{4}}}{3 \times 3^{\frac{1}{3}}} = \frac{5^{\frac{2}{4}} \times 3^{\frac{2}{4} \frac{1}{3}}}{3} = \frac{5^{\frac{1}{2}} \times 3^{\frac{1}{6}}}{3} = \frac{1}{3}\sqrt{5}\sqrt[6]{3}$$

Simplifying Radical Expressions

To simplify a radical expression, follow these steps:

1. Factor the number or coefficient completely.

2. For square roots, group like factors in pairs. For cube roots, arrange like factors in groups of three. For nth roots, group like factors in groups of n.

3. For each of these groups, put one of that number outside the radical. Any factors that cannot be combined in groups should be multiplied together and left inside the radical.

4. The index number of a radical is the little number on the front of the radical. For a cube root, the index is 3. If no index appears, then the index is 2 for square roots.

5. For variables inside the radical, divide the index number of the radical into each exponent. The quotient (the answer to the division) is the new exponent to be written on the variable outside the radical. The remainder from the division is the new exponent on the variable remaining inside the radical sign. If the remainder is zero, then the variable no longer appears in the radical sign.

Note: *Remember that the square root of a negative number can be designated by replacing the negative sign inside that square root with an i in front of the radical (to signify an imaginary number). Then simplify the remaining positive radical by the normal method. Include the i outside the radical as part of the answer. If the index number is an odd number, you can still simplify the radical to get a negative solution.*

Example:
$$\sqrt{50a^4b^7} = \sqrt{5 \times 5 \times 2 \times a^4 \times b^7} = 5a^2b^3\sqrt{2b}$$

Example:
$$7x\sqrt[3]{16x^5} = 7x\sqrt[3]{2 \times 2 \times 2 \times 2 \times x^5} = 7x \times 2x\sqrt[3]{2x^2} = 14x^2\sqrt[3]{2x^2}$$

An expression with a radical sign can be rewritten using a rational exponent. The radicand becomes the base, which will have the rational exponent. The index number on the front of the radical sign becomes the denominator of the rational exponent. The numerator of the rational exponent is the exponent, which was originally inside the radical sign on the original base. *Note*: If no index number appears on the front of the radical, then it is a 2. If no exponent appears inside the radical, then use a 1 as the numerator of the rational exponent.

If an expression contains rational expressions with different denominators, rewrite the exponents with a common denominator and then change the problem into a radical.
$$a^{\frac{2}{3}} b^{\frac{1}{2}} c^{\frac{3}{5}} = a^{\frac{20}{30}} b^{\frac{15}{30}} c^{\frac{18}{30}} = \sqrt[30]{a^{20} b^{15} c^{18}}$$

Solving Radical Equations

Follow these steps to solve a radical equation:

1. Get a term with a radical alone on one side of the equation.

2. Raise both sides of the equation to a power equal to the index on the radical (that is, square both sides of an equation containing a square root, cube both sides of an equation containing a cube root,

Solving Radical Equations

Follow these steps to solve a radical equation:

1. Get a term with a radical alone on one side of the equation.

2. Raise both sides of the equation to a power equal to the index on the radical (that is, square both sides of an equation containing a square root, cube both sides of an equation containing a cube root, etc.). *Do not square (or cube, etc.) each term separately. Square (or cube, etc.) the entire side of the equation.*

3. If there are any radicals remaining, repeat steps 1 and 2 until all radicals are gone.

4. Solve the remaining equation.

5. Check your answers in the original radical equation. Not every answer may check out. If no answer checks out in the original equation, then the answer to the equation is ∅, the empty set or null set.

Solve and check:

$\sqrt{2x-8} - 7 = 9$	Get radical alone.
$\sqrt{2x-8} = 16$	Square both sides.
$\sqrt{2x-8}^2 = 16^2$	
$2x - 8 = 256$	Solve for x.
$2x = 264$	
$x = 132$	

Check:

$\sqrt{2(132) - 8} - 7 = 9$	
$\sqrt{264 - 8} - 7 = 9$	
$\sqrt{256} - 7 = 9$	
$16 - 7 = 9$	This answer checks.

Note: Since the answers to radical equations must be checked in the original equation anyway, this means that possible answers on a multiple choice test could be immediately substituted into the problem to find the correct answer without having to solve the actual problem.

Solve and check:

$\sqrt{5x-1} - 1 = x$	Add 1.
$(\sqrt{5x-1})^2 = (x+1)^2$	Square both sides.
$5x - 1 = x^2 + 2x + 1$	Solve this equation.
$0 = x^2 - 3x + 2$	
$0 = (x-2)(x-1)$	
$x = 2 \qquad x = 1$	

Check both answers:

$$\sqrt{5(2)-1} - 1 = 2 \qquad \sqrt{5(1)-1} - 1 = 1$$
$$3 - 1 = 2 \qquad\qquad 2 - 1 = 1$$

Both answers check.

Comparing Radicals

Radicals of the same order can be compared by considering the numbers under the radical sign. Radicals of different orders can be compared after reducing them to radicals of the same order. When an expression has a rational exponent, it can be rewritten using a radical sign. The denominator of the rational exponent becomes the index number on the front of the radical sign. The base of the original expression goes inside the radical sign. The numerator of the rational exponent is an exponent, which can be placed either inside the radical sign on the original base or outside the radical as an exponent on the radical expression. The radical can then be simplified as far as possible.

$$4^{\frac{3}{2}} = \sqrt[2]{4^3} \text{ or } (\sqrt{4})^3 = \sqrt{64} = 8$$
$$16^{\frac{3}{4}} = \sqrt[4]{16^3} \text{ or } (\sqrt[4]{16})^3 = 2^3 = 8$$
$$25^{\frac{-1}{2}} = \frac{1}{25^{\frac{1}{2}}} = \frac{1}{\sqrt{25}} = \frac{1}{5}$$

Example: Compare $\sqrt{5}$ and $\sqrt[3]{11}$.
$$\sqrt{5} = 5^{\frac{1}{2}} \text{ and } \sqrt[3]{11} = 11^{\frac{1}{3}}$$

Here the powers are $\frac{1}{2}$ and $\frac{1}{3}$. The LCD will be 6.

$$\sqrt{5} = 5^{\frac{1}{2}} = 5^{\frac{3}{6}} = (5^3)^{\frac{1}{6}} = (125)^{\frac{1}{6}} \text{, and}$$
$$\sqrt[3]{11} = 11^{\frac{1}{3}} = 11^{\frac{2}{6}} = (11^2)^{\frac{1}{6}} = 121^{\frac{1}{6}}$$

Now both of the given radicals are of the same power. Therefore, we can compare the bases.

Since $125 > 121$, obviously $\sqrt{5} > \sqrt[3]{11}$.

Example: Arrange the following radicals in ascending order of magnitude:
$$\sqrt[3]{2}, \sqrt[4]{3}, \sqrt[6]{5}$$

$$\sqrt[3]{2} = 2^{\frac{1}{3}} = 2^{\frac{20}{60}} = (2^{20})^{\frac{1}{60}} = (1048576)^{\frac{1}{60}}$$
$$\sqrt[4]{3} = 3^{\frac{1}{4}} = 3^{\frac{15}{60}} = (3^{15})^{\frac{1}{60}} = (14348907)^{\frac{1}{60}}$$
$$\sqrt[6]{5} = 5^{\frac{1}{6}} = 5^{\frac{10}{60}} = (5^{10})^{\frac{1}{60}} = (9765625)^{\frac{1}{60}}$$

It is clear that $1048576 < 9765625 < 14348907$;
$$\therefore \sqrt[3]{2} < \sqrt[6]{5} < \sqrt[4]{3}.$$

The Laws of Exponents

For any real number x and integers m and n

1. $x^m \times x^n = x^{m+n}$

2. $\frac{x^m}{x^n} = x^{m-n}$, if $m > n$

   $\qquad = \frac{1}{x^{n-m}}$, if $n > m$

3. $(x^m)^n = x^{mn}$

4. $(xy)^m = x^m \times y^n$

5. $x^0 = 1$, for all finite values of x

6. $x^{-m} = \frac{1}{x^m}$

Example: Simplify

$$(a^{\frac{1}{3}}b^{-2}c)^3 \times (a^2 b^{\frac{3}{2}} c^{-1})^4 \div (a^2 b^{\frac{1}{4}} c^{\frac{1}{2}}).$$

The given expression is $\dfrac{(a^{\frac{1}{3}} b^{-2} c)^3 \times (a^2 b^{\frac{3}{2}} c^{-1})^4}{(a^2 b^{\frac{1}{4}} c^{\frac{1}{2}})} = \dfrac{a^{\frac{1}{3} \times 3 + 2 \times 4} b^{-2 \times 3 + \frac{3}{2} \times 4} c^{3 + (-4)}}{a^2 b^{\frac{1}{4}} c^{\frac{1}{2}}}$

$\quad = a^{9 - 2} b^{0 - \frac{1}{4}} c^{-1 - \frac{1}{2}}$

$\quad = a^7 b^{\frac{-1}{4}} c^{\frac{-3}{2}}$

SKILL 1.18 **Determine the equations of lines, given sufficient information; recognize and use the basic forms of the equations for a straight line**

Basic Equation Forms For a Straight Line

The standard form for the equation of a straight line is $ax + by = c$ where a, b and c are constants and a and b are not both zero.

The equation of a straight line can also be written in slope-intercept form, $y = mx + b$, where m is the slope and b is the y-intercept.

The slope of a line is the "slant" of the line. A line slanting downward from left to right has a negative slope. A line slanting upward from left to right has a positive slope.

The y-intercept is the y-coordinate of the point where a line crosses the y-axis.

An alternate form of a linear equation is the point-slope form. Given the slope of a line and any one point (x_a, y_a) the line passes through, its equation may be written as

$$y - y_a = m(x - x_a)$$

There are other forms for the equation of a straight line but these are the more commonly used ones. Any of the forms, of course, may be transformed into any of the other forms through algebraic manipulation.

Determining the Equation of a Straight Line

Equation of a line given the slope and y-intercept

Given the slope and y-intercept of a line, its equation can be written by substituting the values of the slope and intercept in the slope-intercept form $y = mx + b$ (m is the slope, b is the y-intercept).

The equation of a graph can be found by finding its slope and its intercept.

To find the slope, find 2 points on the graph where coordinates are integer values. Using points: (x_1, y_1) and (x_2, y_2).

$$\text{slope} = \frac{y_2 - y_1}{x_2 - x_1}$$

The y intercept is the y coordinate of the point where a line crosses the y axis.

Example: The slope of a line is -1. If the line crosses the y-axis at y = 5, what is the equation of the line.

The slope of the line $m = $ -1 and the y-intercept $= 5$. Hence the equation is

$$y = \text{-}x + 5$$

Equation of a line given the slope and one point

The point-slope form is the most convenient form to use when the slope of a line and one point is given.

Example: Find the equation of a line that has a slope of -1.5 and passes through the point (3,2). Express the equation in the slope-intercept and standard forms.

Substituting the values of the slope m and the point (x_a, y_a) in the point-slope form of the equation we get

$$y - 2 = \text{-}1.5(x - 3) \qquad \text{(point-slope form)}$$

Rearranging the terms,

$$y = \text{-}1.5x + 6.5 \qquad \text{(slope-intercept form)}$$

or, multiplying by 2 and moving the x-term to the left hand side,

$$3x + 2y = 13 \qquad \text{(standard form)}$$

Equation of a line that passes through two points

Given two points on a line, the first thing to do is to find the slope of the line.

If 2 points on the graph are (x_1, y_1) and (x_2, y_2), then the slope is found using the formula:

$$\text{slope} = \frac{y_2 - y_1}{x_2 - x_1}$$

The slope will now be denoted by the letter m. To write the equation of the line in point-slope form, choose either point. Substitute it into the formula:

$$y - y_a = m(x - x_a)$$

Remember (x_a, y_a) can be (x_1, y_1) or (x_2, y_2). If m, the value of the slope, is distributed through the parentheses, the equation can be rewritten into other forms of the equation of a line.

Alternatively, the given point can also be used to find the *y*-intercept of the line which is then used to write the equation for the line in the slope-intercept form.

Example: Find the equation of a line that passes through the points (9,-6) and (-1, 2) in standard form.

$$\text{slope} = \frac{y_2 - y_1}{x_2 - x_1} = \frac{2 - (\text{-}6)}{\text{-}1 - 9} = \frac{8}{\text{-}10} = -\frac{4}{5}$$

The *y*-intercept may be found by substituting the slope (*m*) and the coordinates (*x, y*) for one of the data points in the slope-intercept form of the equation *y* = *mx* + *b* giving

$$\text{-}6 = -\frac{4}{5} \times 9 + b \quad \text{where } b \text{ is the } y\text{-intercept}$$

$$b = \frac{6}{5}$$

Thus the slope-intercept form of the equation is:

$$y = -\frac{4}{5}x + \frac{6}{5}$$

Multiplying by 5 to eliminate fractions, it is:

$$5y = \text{-}4x + 6 \;\rightarrow\; 4x + 5y = 6 \qquad \text{Standard form}$$

Equation of a line that is perpendicular or parallel to a given line

Lines that are equidistant from each other and never intersect are parallel. Parallel lines, therefore, must have the same slope. Note that vertical lines are parallel to other vertical lines despite having undefined slopes.

Perpendicular lines form 90º angles at their intersection. The slopes of perpendicular lines are the negative reciprocals of each other. In other words, if a line has a slope of -2, a perpendicular line would have a slope of $\frac{1}{2}$. Note that the product of these slopes is -1.

Example: Find the equation of the line that passes through the point (2, 1) and is perpendicular to the line y = 2x − 3.
The slope of the line $m = -\frac{1}{2}$.

In the point slope form:

$$(y - 1) = \text{-}0.5\,(x - 2)$$

Rearranging,

$$y = \text{-}0.5\,x + 2 \qquad \text{or} \qquad 2y + x = 4$$

Solve and graph linear equations and inequalities in one or two variables; solve and graph systems of linear equations and inequalities in two variables

Graphing Linear Inequalities

INEQUALITY: a mathematical expression containing the symbol >, <, ≥, or ≤

To graph an INEQUALITY, solve the inequality for y. This gets the inequality in SLOPE INTERCEPT FORM—for example: $y < mx + b$. The point $(0, b)$ is the y-intercept and m is the line's slope.

- If the inequality solves to $x >$, $\geq$, $<$, or $\leq$ any number, then the graph includes a *vertical line*.

- If the inequality solves to $y >$, $\geq$, $<$, or $\leq$ any number, then the graph includes a *horizontal line*.

SLOPE INTERCEPT FORM: a line is $y = mx + b$, where m is the slope of the line and $(0, b)$ is the y-intercept.

When graphing a linear inequality, the line will be dotted if the inequality sign is $<$ or $>$. If the inequality sign is either $\geq$ or $\leq$, the line on the graph will be a solid line. Shade above the line when the inequality sign is $\geq$ or $>$. Shade below the line when the inequality sign is $\leq$ or $<$. For inequalities of the forms $x >$ number, $x \leq$ number, $x <$ number, or $x \geq$ number, draw a vertical line (solid or dotted). Shade to the right for $>$ or $\geq$. Shade to the left for $<$ or $\leq$.

Remember: Dividing or multiplying by a negative number will reverse the direction of the inequality sign.

Remember: Dividing or multiplying by a negative number will reverse the direction of the inequality sign.

Use these rules to graph and shade each inequality. The solution to a system of linear inequalities consists of the part of the graph where the shaded areas for all the inequalities in the system overlap. For instance, if the graph of one inequality was shaded with red, and the graph of another inequality was shaded with blue, then the overlapping area would be shaded with purple. The points in the purple area would be the solution set of this system.

Example: Solve by graphing:

$$x + y \leq 6$$
$$x - 2y \leq 6$$

Solving the inequalities for y, we find that they become:

$$y \leq -x + 6 \ (y\text{-intercept of 6 and slope} = -1)$$
$$y \geq \frac{1}{2}x - 3 \ (y\text{-intercept of -3 and slope} = \frac{1}{2})$$

A graph with shading is shown below:

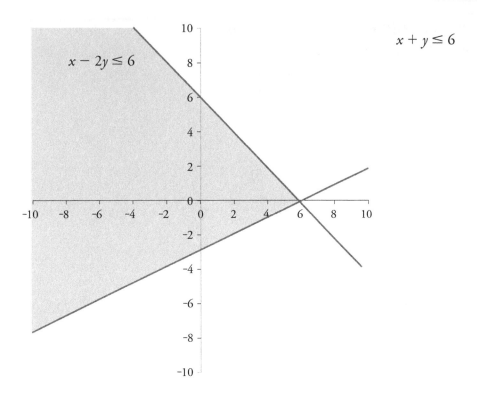

Solving Equations and Inequalities

To solve an equation or inequality, follow these steps:

Step 1	If there are parentheses, use the distributive property to eliminate them.
Step 2	If there are fractions, determine their LCD (least common denominator). Multiply every term of the equation by the LCD. This will cancel out all of the fractions, making it easier to solve the equation or inequality.
Step 3	If there are decimals, find the largest decimal. Multiply each term by a power of 10 (10, 100, 1,000, etc.) with the same number of zeros as the length of the decimal. This will eliminate all decimals while solving the equation or inequality.
Step 4	Combine the like terms on each side of the equation or inequality.
Step 5	If there are variables on both sides of the equation, add or subtract one of those variable terms to move it to the other side. Combine like terms.
Step 6	If there are constants on both sides, add or subtract one or more of those constants to both sides. Combine like terms.
Step 7	If there is a coefficient in front of the variable, divide both sides by this number to get the answer to the equation. Remember: *Dividing or multiplying an inequality by a negative number will reverse the direction of the inequality sign.*
Step 8	The solution of a linear equation is a single number. The solution of an inequality is a range of values shown using an inequality sign.

Example: Solve 3(2x + 5) − 4x = 5(x + 9).

$6x + 15 − 4x = 5x + 45$	Step 1
$2x + 15 = 5x + 45$	Step 4
$-3x + 15 = 45$	Step 5
$-3x = 30$	Step 6
$x = -10$	Step 7

Example: Solve $\frac{1}{2}(5x + 34) = \frac{1}{4}(3x − 5)$.

$\frac{5}{2}x + 17 = \frac{3}{4}x − \frac{5}{4}$	Step 1

The LCD of $\frac{5}{2}, \frac{3}{4},$ and $\frac{5}{4}$ is 4.
Multiply by the LCD of 4.

$4(\frac{5}{2}x + 17) = (\frac{3}{4}x − \frac{5}{4})4$	Step 2
$10x + 68 = 3x − 5$	
$7x + 68 = -5$	Step 5
$7x = -73$	Step 6
$x = \frac{-73}{7}$ or $-10\frac{3}{7}$	Step 7

Check:

$$\frac{1}{2}[5\frac{-73}{7} + 34] = \frac{1}{4}[3(\frac{-73}{7}) − 5]$$

$$\frac{-73(5)}{14} + 17 = \frac{3(-73)}{28} − \frac{5}{4}$$

$$\frac{-73(5) + 17(14)}{14} = \frac{3(-73)}{28} - \frac{5}{4}$$

$$\frac{-73(5) + 17(14)}{14} = \frac{3(-73) − 35}{28}$$

$$(\frac{-365 + 238}{28}) \times 2 = \frac{-219 − 35}{28}$$

$$\frac{-254}{28} = \frac{-254}{28}$$

Example: Solve 6x + 21 < 8x + 31.

$-2x + 21 < 31$	Step 5
$-2x < 10$	Step 6
$x > -5$	Step 7

Note that the inequality sign has changed direction.

Optimization Problems

LINEAR PROGRAMMING and inequalities can be used to solve various types of word problems and can be used in a practical way to solve real-world problems. It is often used in various industries, ecological sciences, and governmental organizations to determine or project production costs, the amount of pollutants dispersed into the air, and so forth. The key to most linear programming problems is to organize the information in the word problem into a chart or graph of some type.

LINEAR PROGRAM-MING: a mathematical technique used to optimize a linear objective function that is subject to certain constraints

Example: A printing manufacturer makes two types of printers: a Printmaster and a Speedmaster. The Printmaster takes up 10 cubic feet of space and weighs 5,000 pounds, and the Speedmaster takes up 5 cubic feet of space and weighs 600 pounds. The total available space for storage before shipping is 2,000 cubic feet, and the weight limit for the space is 300,000 pounds. The profit on the Printmaster is $125,000, and the profit on the Speedmaster is $30,000. How many of each machine should be sold to maximize profitability, and what is the maximum possible profit?

First, let x represent the number of Printmasters sold and y represent the number of Speedmasters sold. Therefore, the equation for space is $10x + 5y \leq 2,000$, which simplifies to $2x + y \leq 400$, which in terms of y is $y \leq -2x + 400$. Since you can't have a negative number of printers, $x \geq 0$ and $y \geq 0$. The equation for weight is $5,000x + 600y \leq 300,000$, which simplifies to $50x + 6y \leq 3,000$.

Substitute $x = 0$ and $y = 0$ in both equations to find where each linear equation meets the x- and y-axes, and you get the points (0, 400) and (200, 0) for space and the points (60, 0) and (0, 500) for weight.

By using substitution, we can solve the system of equations for space and weight:

$50x + 6(-2x + 400) \leq 3000$

$50x - 12x + 2400 \leq 3000$

$38x \leq 600$

$x \leq 15.8$ Substitute 15.8 back into one of the equations for x to solve for y:

$2(15.8) + y \leq 400$

$31.6 + y \leq 400$

$y \leq 368.4$

The equation for profit is $125,000x + 30,000y$. To find the maximum values, make a table using the information you found (numbers found above were rounded to whole numbers):

Vertexes	125,000x	30,000y	125,000x + 30,000y
(0,0)	0	0	0
(0,400)	0	12,000,000	12,000,000
(16,368)	2,000,000	11,040,000	13,040,000
(60,0)	7,500,000	0	7,500,000

You can see from the table that the maximum profit is reached when 16 Printmasters (x) are sold and 368 Speedmasters (y) are sold, to get a maximum profit of $13,040,000.

This can be shown visually in a graph. Here is a graph of the linear inequality for space and for weight and their point of intersection, (16, 368), which tells you the number of each printer that needs to be sold for maximum profitability based on the weight and space constraints.

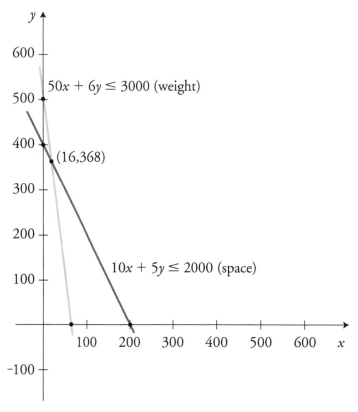

Example: The YMCA volunteers want to sell raffle tickets to raise at least $32,000. If they must pay $7,250 in expenses and prizes out of the money collected from the tickets, how many tickets worth $25 each must they sell?
Solution: Since they want to raise *at least* $32,000, that means they would be happy to get $32,000 *or more*. This requires an inequality.

Let x = number of tickets sold

Then $25x$ = total money collected for x tickets

Total money minus expenses is greater than $32,000.

$25x - 7250 \geq 32000$

$25x \geq 39250$

$x \geq 1570$

If they sell 1,570 tickets or more, they will raise *at least* $32,000.

Example: Sharon's Bike Shoppe can assemble a 3-speed bike in 30 minutes or a 10-speed bike in 60 minutes. The profit on each bike sold is $60 for a 3-speed or $75 for a 10-speed bike. How many of each type of bike should they assemble during an eight-hour day (480 minutes) to make the maximum profit? Total daily profit must be at least $300.

  Let x = number of 3-speed bikes

    y = number of 10-speed bikes

Since there are only 480 minutes to use each day, $30x + 60y \leq 480$ is the first inequality.

Since the total daily profit must be at least $300, $60x + 75y \geq 300$ is the second inequality.

  $30x + 60y \leq 480$ simplifies to $y \leq 8 - \frac{x}{2}$

  $60y \leq -30x + 480$

   $y \leq -\frac{1}{2}x + 8$

  $60x + 75y \geq 300$ simplifies to $y \geq 4 - \frac{4x}{5}$

  $75y + 60x \geq 300$

    $y \geq -\frac{4x}{5} + 4$

Graph these two inequalities:

  $y \leq 8 - \frac{x}{2}$

  $y \geq 4 - \frac{4x}{5}$

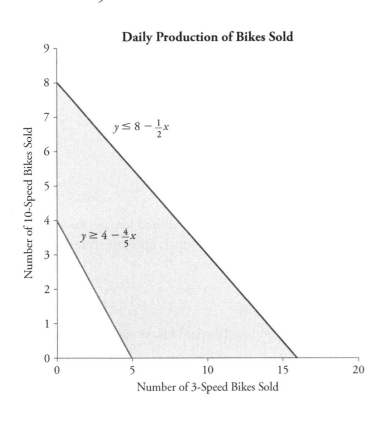

Daily Production of Bikes Sold

$y \leq 8 - \frac{1}{2}x$

$y \geq 4 - \frac{4}{5}x$

Number of 10-Speed Bikes Sold

Number of 3-Speed Bikes Sold

Realize that $x \geq 0$ and $y \geq 0$, since the number of bikes assembled cannot be a negative number. Graph these as additional constraints on the problem. The number of bikes assembled must always be an integer value, so points within the shaded area of the graph must have integer values. The maximum profit will occur at or near a corner of the shaded portion of this graph. Those points occur at $(0, 4)$, $(0, 8)$, $(16, 0)$, or $(5, 0)$.

Since profits are \$60/three-speed and \$75/ten-speed, the profits for these four points would be:

$(0, 4)$	$60(0) + 75(4) = 300$
$(0, 8)$	$60(0) + 75(8) = 600$
$(16, 0)$	$60(16) + 75(0) = 960 \leftarrow$ Maximum profit
$(5, 0)$	$60(5) + 75(0) = 300$

The maximum profit would occur if 16 3-speed bikes were made daily. The profit curve is added as shown here. (It does not show exact numbers, just the trend line showing how profit changes along the upper border of the shaded area.)

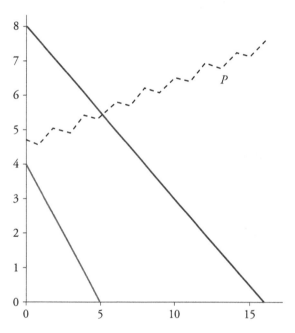

Based on the above plot, it is clear that the profit is maximized for the case where only 3-speed bikes (corresponding to x) are manufactured. Thus, the correct solution can be found by solving the first inequality for y = 0:

$$x + 2(0) \leq 16$$
$$x \leq 16$$

The manufacture of 16 3-speed bikes (and no 10-speed bikes) maximizes profit to \$960 per day.

Matrices can also be used to solve systems of linear inequalities. Please see Domain IV B, Skill 4.5, for more information.

Polynomials

To understand the theorems that follow, you must have some knowledge of POLYNOMIALS. You may have dealt with expressions like $2x^2$ or $5x$ or $6x$. Polynomials can be a single expression or sums of these expressions, and each part of the polynomial is called a term. The terms can have variables in them and exponents, but there are no fractional exponents, no square roots of exponents, and no variables in the denominator.

Terms

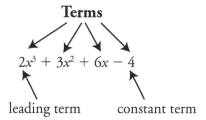

$$2x^3 + 3x^2 + 6x - 4$$

leading term   constant term

The constant term has no variable and therefore never changes. It is the y-intercept.

Polynomials written in standard form have the terms written in decreasing value according to the value of the exponent, like above. This particular polynomial is a "degree 3" polynomial because the exponent of the first term is 3.

A special case of a polynomial is the quadratic equation, which is of degree 2: $ax^2 + bx + c = 0$. There are many ways to solve for x in polynomials and quadratics, but sometimes the only way to solve for x in a quadratic equation is to use the quadratic formula:

$$x = \frac{-b \pm \sqrt{b^2 - 4ac}}{2a}$$

The Rational Root Theorem

The RATIONAL ROOT THEOREM theorem is also known as the Rational Zero Theorem and offers a list of all possible rational roots or zeros of the polynomial equation (all coefficients are integers):

$$a_n x^n + a_{n-1} x^{n-1} + a_{n-2} x^{n-2} \ldots a_2 x^2 + a_1 x + a_0 = 0$$

POLYNOMIALS: these can be a single expression or sums of these expressions, and each part of the polynomial is called a term; the terms can have variables in them and exponents, but there are no fractional exponents, no square roots of exponents, and no variables in the denominator

RATIONAL ROOT THEOREM: also known as the Rational Zero Theorem, this offers a list of all possible rational roots or zeros of the polynomial equation

A root or a zero is a solution for x in the polynomial equation and is where the graph of the polynomial crosses the x-axis and the y-value is zero.

Every rational solution, zero or root of x can be written as $x = \pm\frac{p}{q}$,

where p is an integer factor of the constant term a_0 and q is an integer factor of the leading coefficient a_n.

Example: $2x^2 + x - 3 = 0$ and $x = \pm\frac{p}{q} = \pm\frac{1,3}{1,2}$, which results in possible answers of $1, -1, 3, -3, \frac{1}{2}, -\frac{1}{2}, \frac{3}{2}, -\frac{3}{2}$. Since this is a quadratic equation of degree 2, there are only two roots or zeros.

The Factor Theorem

The FACTOR THEOREM establishes the relationship between the factors and the zeros or roots of a polynomial and is useful for finding the factors of higher-degree polynomials. It states that a polynomial $f(x)$ has a factor $x - a$ if and only if $f(a) = 0$.

Example: Find the factors of the polynomial $x^3 + 2x^2 - x - 2$.

We use trial and error to find the first factor, and when we find one and substitute it into the preceding equation, and the equation is equal to zero, we know we have found a factor. Is $(x - 2)$ a factor? To find out, substitute 2 into the preceding equation:

$2^3 + 2(2)^2 - 2 - 2 = 8 + 8 - 4 = 12$. We now know that $(x - 2)$ is not a factor because when we substituted in 2 for x, we got 12 as a solution, not zero.

Let's try the polynomial factor $(x + 2)$ or $x = -2$:

$(-2)^3 + 2(-2)^2 - (-2) - 2 = -8 + 8 + 2 - 2 = 0$. The equation is equal to zero, so we now know we have a factor and $x = -2$ is a root or zero. To find the remaining roots, we can divide our original polynomial by the factor we found: $\frac{x^3 + 2x^2 - x - 2}{x + 2} = x^2 - 1$, which factors to: $(x + 1)(x - 1)$. We now have the three factors of $x^3 + 2x^2 - x - 2$: $(x + 2)$, $(x + 1)$, and $(x - 1)$. We also have the roots or zeros of $x^3 + 2x^2 - x - 2$: $(-2, -1, 1)$.

The Complex Conjugate Root Theorem

The COMPLEX CONJUGATE ROOT THEOREM states that if P is a polynomial function with real number coefficients and $a + bi$ (b is not zero) is a root of $P(x) = 0$, then $a - bi$ is also a root of P.

It follows that since complex factors come in pairs (an even number), when these factors are multiplied, the product is a quadratic polynomial with real coefficients.

Example: Find the complex roots of $4x^3 + 15x - 36 = 0$, which is already in the standard form.

The equation is in order with powers of x from highest to lowest. Note that there are no common factors. The equation is degree 3, so there will be three roots or zeros. There is one variation in sign, meaning that there will be one real root according to Descartes's Rule of Signs. According to the Rational Root Theorem, $x = \pm\frac{p}{q}$, we have numerous possibilities for roots as the numbers 1, 2, 3, 4, 6, 9, 12, 18, and 36 (factors of the constant term, p), are divided by 4, 2, and 1 (factors of the first coefficient q). The other two roots are a pair of complex conjugates. So let's start with $x = +\frac{p}{q} = \frac{1}{1}$ or 1.

In order to determine whether $x = 1$ is a root of the given polynomial, we can use synthetic division. This process is described in the next section.

Synthetic Division

SYNTHETIC DIVISION is used to divide a polynomial by a binomial of the form $x - a$. This is often done in order to find the value of a function at $x = a$ since, according to the Remainder Theorem, $f(a)$ is equal to remainder when $f(x)$ is divided by $(x - a)$. Synthetic division is also used to identify the roots of a polynomial since the remainder of division of $f(x)$ by $(x - a)$ is zero when $x = a$ is a root of $f(x)$.

> **SYNTHETIC DIVISION:** used to divide a polynomial by a binomial of the form $x - a$

To perform synthetic division, divide the value of x into the coefficients of the function (remember that coefficients of missing terms, like x^2 below, must be included).

Continuing with the example introduced in the previous section, we use synthetic division to divide $4x^3 + 15x - 36$ by $(x - 1)$ as shown below:

$$
\begin{array}{r|rrrr}
1 & 4 & 0 & 15 & -36 \\
  &   & 4 & 4 & 19 \\
\hline
  & 4 & 4 & 19 & -17
\end{array}
$$
$\leftarrow$ The value of the function

We did not get zero as a remainder, so 1 is not a root of the polynomial $4x^3 + 15x - 36$. The result also tells us that

$(4x^3 + 15x - 36) \div (x - 1) = 4x^2 + 4x + 19$ with remainder -17

Let's try division by $x - 2$:

```
   2 | 4   0   15   -36
     |     8   16    62
     ‾‾‾‾‾‾‾‾‾‾‾‾‾‾‾‾‾‾‾‾
       4   8   31    26  ← The value of the function
```

We didn't get zero as a remainder, so 2 is not a root, but $f(1) = -17$ and $f(2) = 26$, so the zero or root lies somewhere between $x = 1$ and $x = 2$, and the only possible root we could choose from the preceding list of possibilities would be $\frac{3}{2}$:

```
  3/2 | 4   0   15   -36
      |     6    9    36
      ‾‾‾‾‾‾‾‾‾‾‾‾‾‾‾‾‾‾‾
        4   6   24     0  ← The value of the function
```

We found that $\frac{3}{2}$ works and is a root, since the remainder is 0.

Next, we divide our original polynomial by $(x - \frac{3}{2})$ to reduce the polynomial to a second-degree equation, which is a quadratic:

$$\frac{4x^3 + 15x - 36}{x - \frac{3}{2}} = 2x^2 + 3x + 12 = 0.$$

We know from Descartes's Rule of Signs that there is only one real root, so there is no sense in trying to factor this quadratic equation. We will have to use the quadratic formula to find the complex roots

$$x = \frac{-b \pm \sqrt{b^2 - 4ac}}{2a}$$

$$x = \frac{-3 \pm \sqrt{9 - 4(2)(12)}}{2(2)} = \frac{-3 \pm \sqrt{-87}}{4} = -\frac{3}{4} + \left(\frac{\sqrt{-87}}{4}\right)i$$

and $-\frac{3}{4} - \left(\frac{\sqrt{-87}}{4}\right)i$

Example: Find the roots of $x^2 - 6x + 13 = 0$

Using the quadratic formula, we substitute in and find:

$$x = \frac{6 \pm \sqrt{36 - 4(13)}}{2} = \frac{6 \pm \sqrt{-16}}{2} = \frac{6 \pm \sqrt{-1}\sqrt{16}}{2} = \frac{6 \pm 4i}{2} = 3 + 2i, 3 - 2i$$

In both cases, we found the pair of complex numbers that are factors of the preceding polynomials. Any polynomial equation of degree n has exactly n zeros if we allow complex numbers, and we can solve previously unsolvable polynomials by simply defining the $\sqrt{-1} = i$.

The Quadratic Formula

QUADRATIC EQUATION: an equation in which one or more of the terms is squared but raised to no higher power

A QUADRATIC EQUATION is written in the form $ax^2 + bx + c = 0$. One method of solving it is by factoring the quadratic expression and applying the condition that at least one of the factors must equal zero in order for the whole expression to be zero.

Example: Solve the equation.

$x^2 + 10x - 24 = 0$

$(x + 12)(x - 2) = 0$ Factor.

$x + 12 = 0$ or $x - 2 = 0$ Set each factor equal to 0.

$x = -12$ $x = 2$ Solve.

Check:

$x^2 + 10x - 24 = 0$

$(-12)^2 + 10(-12) - 24 = 0$ $(2)^2 + 10(2) - 24 = 0$

$144 - 120 - 24 = 0$ $4 + 20 - 24 = 0$

$0 = 0$ $0 = 0$

The quadratic formula proof

$ax^2 + bx + c = 0$

$x^2 + \frac{b}{a}x + \frac{c}{a} = 0$

$x^2 + \frac{b}{a}x = -\frac{c}{a}$

$x^2 + \frac{b}{a}x + \left(\frac{b}{2a}\right)^2 = -\frac{c}{a} + \left(\frac{b}{2a}\right)^2$ (Complete the square.)

$x^2 + \frac{b}{a}x + \frac{b^2}{4a^2} = -\frac{c}{a} + \frac{b^2}{4a^2}$

$\left(x + \frac{b}{2a}\right)\left(x + \frac{b}{2a}\right) = -\frac{c}{a}\left(\frac{4a}{4a}\right) + \frac{b^2}{4a^2}$ (Factor left side; use LCD on right.)

$\left(x + \frac{b}{2a}\right)^2 = \frac{b^2 - 4ac}{4a^2}$

$\sqrt{\left(x + \frac{b}{2a}\right)^2} = \pm\sqrt{\frac{b^2 - 4ac}{4a^2}}$

$x + \frac{b}{2a} = \pm\frac{\sqrt{b^2 - 4ac}}{2a}$ Now solve for x:

$x + \frac{b}{2a} - \frac{b}{2a} = \pm\frac{\sqrt{b^2 - 4ac}}{2a} - \frac{b}{2a}$

$x = \frac{-b \pm \sqrt{b^2 - 4ac}}{2a}$ and you now have the Quadratic Formula.

FEATURES OF A PARABOLA		
Form of Equation	$y = a(x - h)^2 + k$	$x = a(y - k)^2 + k$
Identification	x^2 term, y not squared	y^2 term, x not squared

Table continued on next page

Sketch of Graph		
Axis of Symmetry	(A line through the vertex and focus about which the parabola is symmetric)	
Directrix	$y = k - \frac{1}{4a}$	$x = h - \frac{1}{4a}$
Vertex	(h,k)	(h,k)
Focus	$(h, k + \frac{1}{4a})$	$(h + \frac{1}{4a}, k)$
Direction of Opening	up if $a > 0$, down if $a < 0$	right if $a > 0$, left if $a < 0$
Length of Latus Rectum	$\left\|\frac{1}{a}\right\|$	$\left\|\frac{1}{a}\right\|$

Example: Find all of the identifying features of $y = -3x^2 + 6x - 1$.

First, the equation must be put into the general form $y = a(x - h)^2 + k$.

$y = -3x^2 + 6x - 1$ ⟶ Begin by completing the square.

$\quad = -3(x^2 - 2x + 1) - 1 + 3$

$\quad = -3(x - 1)^2 + 2$ ⟶ Using the general form of the equation, begin to identify known variables.

$a = -3 \quad h = 1 \quad k = 2$

axis of symmetry: $x = 1$

vertex: $(1,2)$

focus: $(1, 1\frac{11}{12})$

directrix: $y = 2\frac{1}{12}$

direction of opening: down since $a < 0$

length of latus rectum: $\frac{1}{3}$

Example: Graph $y = 3x^2 + x - 2$.

x	$y = 3x^2 + x - 2$
-2	8
-1	0
0	-2
1	2
2	12

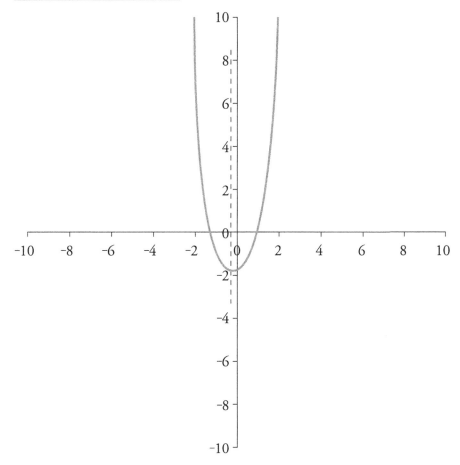

To solve a quadratic equation using the quadratic formula, make sure that your equation is in the form $ax^2 + bx + c = 0$. Substitute these values into the formula:

$$x = \frac{-b \pm \sqrt{b^2 - 4ac}}{2a}$$

Simplify the result to find the answers. (Remember: There could be two real answers, one real answer, or two complex answers that include i.)

Example: Solve the equation.

$3x^2 = 7 + 2x$

$a = 3 \qquad b = -2 \qquad c = -7$

$x = \dfrac{-(-2) \pm \sqrt{(-2)^2 - 4(3)(-7)}}{2(3)}$

$x = \dfrac{2 \pm \sqrt{4 + 84}}{6}$

$x = \dfrac{2 \pm \sqrt{88}}{6}$

$x = \dfrac{2 \pm 2\sqrt{22}}{6}$

$x = \dfrac{1 \pm \sqrt{22}}{3}$

Completing the Square

A quadratic equation can also be solved by completing the square.

Example: Solve the equation.

$x^2 - 6x + 8 = 0$	
$x^2 - 6x = -8$	Move the constant to the right side.
$x^2 - 6x + 9 = -8 + 9$	Add the square of half the coefficient of x to both sides.
$(x - 3)^2 = 1$	Write the left side as a perfect square.
$x - 3 = \pm\sqrt{1}$	Take the square root of both sides.
$x - 3 = 1 \qquad x - 3 = -1$	Solve.
$x = 4 \qquad\qquad x = 2$	

Check:

$x^2 - 6x + 8 = 0$

$4^2 - 6(4) + 8 = 0 \qquad\qquad 2^2 - 6(2) + 8 = 0$

$16 - 24 + 8 = 0 \qquad\qquad 4 - 12 + 8 = 0$

$0 = 0 \qquad\qquad\qquad\quad 0 = 0$

Graphing Quadratic Equations and Inequalities

The general technique for graphing quadratics is the same as for graphing linear equations. Graphing a quadratic equation, however, results in a parabola instead of a straight line.

Example: Solve and graph: $y > x^2 + 4x - 5$.

The axis of symmetry is located at $x = \dfrac{-b}{2a}$. Substituting 4 for b and 1 for a, this formula becomes

$x = \dfrac{-(4)}{2(1)} = \dfrac{-4}{2} = -2$

Find the coordinates of the points on each side of $x = -2$.

x	y
-5	0
-4	-5
-3	-8
-2	-9
-1	-8
0	-5
1	0

Graph these points to form a parabola. Draw out as a dotted line. Since a greater than sign is used, shade above and inside the parabola.

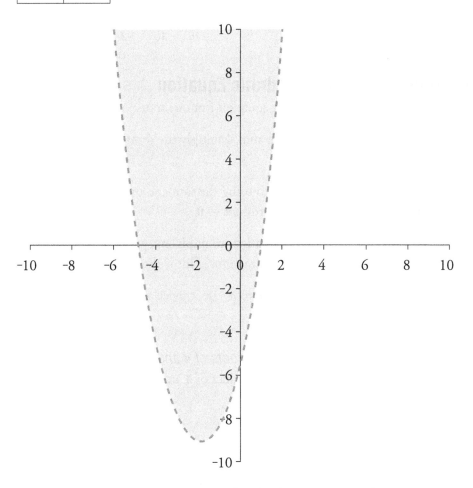

To solve a quadratic inequality with x^2, solve for y. The axis of symmetry is located at $x = \frac{-b}{2a}$. Find the coordinates of the points on each side of the axis of symmetry. Graph the parabola as a dotted line if the inequality sign is either $<$ or $>$. Graph the parabola as a solid line if the inequality sign is either $\leq$ or $\geq$.

Shade above the parabola if the sign is $\geq$ or $>$. Shade below the parabola if the sign is $\leq$ or $<$.

Example: Solve: $8x^2 - 10x - 3 = 0$.
In this equation $a = 8$, $b = -10$, and $c = -3$.

Substituting these into the quadratic equation, it becomes

$$x = \frac{-(-10) \pm \sqrt{(-10)^2 - 4(8)(-3)}}{2(8)} = \frac{10 \pm \sqrt{100 + 96}}{16}$$

$$x = \frac{10 \pm \sqrt{196}}{16} = \frac{10 \pm 14}{16} = \frac{24}{16} = \frac{3}{2} \text{ or } \frac{-4}{16} = \frac{-1}{4}$$

Check:

$x = -\frac{1}{4}$

$\frac{1}{2} + \frac{10}{4} - 3 = 0$ Both check.

$3 - 3 = 0$

$x = \frac{3}{2}$

$8\left(\frac{3}{2}\right)^2 - 10\left(\frac{3}{2}\right) - 3 = 0$

$18 - 15 - 3 = 0$

$3 - 3 = 0$

Using Roots to Write a Quadratic Equation

Follow these steps to write a quadratic equation from its roots:

1. Add the roots together to get their sum. Multiply the roots together to get their *product*.

2. A quadratic equation can be written using the sum and product like this:
   x^2 + (opposite of the sum)x + product $= 0$

3. If there are any fractions in the equation, multiply every term by the common denominator to eliminate the fractions. This is the quadratic equation.

4. If a quadratic equation has only one root, use it twice and follow the first three steps.

Example: Find a quadratic equation with roots of 4 and -9.
Solution: The sum of 4 and -9 is -5. The product of 4 and -9 is -36. The equation would be

   x^2 + (opposite of the sum)x + product $= 0$
   $x^2 + 5x - 36 = 0$

Example: Find a quadratic equation with roots of 5 + 2i and 5 − 2i.
Solution: The sum of $5 + 2i$ and $5 - 2i$ is 10. The product of $5 + 2i$ and $5 - 2i$ is $25 - 4i^2 = 25 + 4 = 29$.

The equation would be

$x^2 + $ **(opposite of the sum)**$x + $ **product** $= 0$

$x^2 - 10x + 29 = 0$

Example: Find a quadratic equation with roots of $\frac{2}{3}$ and $\frac{-3}{4}$.

Solution: The sum of $\frac{2}{3}$ and $\frac{-3}{4}$ is $\frac{-1}{12}$. The product of $\frac{2}{3}$ and $\frac{-3}{4}$ is $\frac{-1}{2}$.

The equation would be

$x^2 + $ **(opposite of the sum)**$x + $ **product** $= 0$

$x^2 + \frac{1}{12}x - \frac{1}{2} = 0$

The common denominator $= 12$, so multiply by 12.

$12x^2 + 1x - 6 = 0$

$12x^2 + x - 6 = 0$

Practice Set Problems

1. Find a quadratic equation with a root of 5

2. Find a quadratic equation with roots of $\frac{8}{5}$ and $\frac{-6}{5}$

3. Find a quadratic equation with roots of 12 and -3

Answer Key

1. $x^2 - 10x + 25 = 0$

2. $25x^2 - 10x - 48 = 0$

3. $x^2 - 9x - 36 = 0$

Factoring the Sum or Difference of Cubes

To factor the sum or the difference of perfect cubes, follow this procedure:

1. Factor out any greatest common factor (GCF).

2. Make a parentheses for a binomial (two terms) followed by a trinomial (three terms).

3. The sign in the first parentheses is the same as the sign in the problem. The difference of cubes will have a "$-$" sign in the first parentheses. The sum of cubes will use a "$+$."

4. The first sign in the second parentheses is the opposite of the sign in the first parentheses. The second sign in the other parentheses is always a "+."

5. Determine what would be cubed to equal each term of the problem. Put those expressions in the first parentheses.

6. To make the three terms of the trinomial, think square-product-square:

   Looking at the binomial, square the first term. This is the trinomial's first term. Looking at the binomial, find the product of the two terms, ignoring the signs. This is the trinomial's second term. Looking at the binomial, square the third term. This is the trinomial's third term. Except in rare instances, the trinomial does not factor again.

Example: Factor completely:

$16x^3 + 54y^3$

$2(8x^3 + 27y^3)$ ← GCF

$2(\quad + \quad)(\quad - \quad + \quad)$ ← signs

$2(2x + 3y)(\quad - \quad + \quad)$ ← what is cube to equal $8x^3$ or $27y^3$

$2(2x + 3y)(4x^2 - 6xy + 9y^2)$ ← square-product-square

$64a^3 - 125b^3$

$(\quad - \quad)(\quad + \quad + \quad)$ ← signs

$(4a - 5b)(\quad + \quad + \quad)$ ← what is cube to equal $64a^3$ or $125b^3$

$(4a - 5b)(16a^2 + 20ab + 25b^2)$ ← square-product-square

$27x^{27} + 343y^{12} = (3x^9 + 7y^4)(9x^{18} - 21x^9y4 + 49y^8)$

Note: The coefficient 27 is different from the exponent 27.

Factoring a Polynomial

To factor a polynomial, follow these steps:

1. Factor out any GCF (greatest common factor).

2. For a binomial (two terms), check to see if the problem is the difference of perfect squares. If both factors are perfect squares, then the binomial factors this way:
   $a^2 - b^2 = (a - b)(a + b)$

3. If the problem is not the difference of perfect squares, then check to see if the problem is either the sum or difference of perfect cubes.
   $x^3 - 8y^3 = (x - 2y)(x^2 + 2xy + 4y^2)$ ← difference
   $64a^3 + 27b^3 = (4a + 3b)(16a^2 - 12ab + 9b^2)$ ← sum
   Note: The sum of perfect squares does not factor.

Note: The sum of perfect squares does not factor.

4. Trinomials could be perfect squares. Trinomials can be factored into two binomials (un-FOILing). Be sure the terms of the trinomial are in descending order. If the last sign of the trinomial is a "$+$," then the signs in the parentheses will be the same as the sign in front of the second term of the trinomial. If the last sign of the trinomial is a "$-$," then there will be one "$+$" and one "$-$" in the two parentheses. The first term of the trinomial can be factored to equal the first terms of the two factors. The last term of the trinomial can be factored to equal the last terms of the two factors. Work backward to determine the correct factors to multiply together to get the correct center term.

Examples:

1. $4x^2 - 25y^2$

2. $6b^2 - 2b - 8$

3. Find a factor of $6x^2 - 5x - 4$

   A. $(3x + 2)$

   B. $(3x - 2)$

   C. $(6x - 1)$

   D. $(2x + 1)$

Answers:

1. No GCF; this is the difference of perfect squares.
   $4x^2 - 25y^2 = (2x - 5y)(2x + 5y)$

2. GCF of 2; Try to factor into two binomials:
   $6b^2 - 2b - 8 = 2(3b^2 - b - 4)$
   Signs are one "$+$," and one "$-$." $3b^2$ factors into $3b$ and b. Find the factors of 4: 1 and 4; 2 and 2.
   $6b^2 - 2b - 8 = 2(3b^2 - b - 4) = 2(3b - 4)(b + 1)$

3. If an answer choice is correct, find the other factor:

   A. $(3x + 2)(2x - 2) = 6x^2 - 2x - 4$

   B. $(3x - 2)(2x + 2) = 6x^2 + 2x - 4$

   C. $(6x - 1)(x + 4) = 6x^2 + 23x - 4$

   D. $(2x + 1)(3x - 4) = 6x^2 - 5x - 4 \leftarrow$ correct factors

Binomial theorem

The BINOMIAL EXPANSION THEOREM is a method used to find the coefficients of $(x + y)^n$. Although Pascal's Triangle is easy to use for small values of n, it can become cumbersome to use with larger values of n.

> **BINOMIAL EXPANSION THEOREM:** a method used to find the coefficients of $(x + y)^n$

The Binomial Theorem states that for any positive value of n,

$$(x + y)^n = x^n + \frac{n!}{(n-1)!1!}x^{n-1}y + \frac{n!}{(n-2)!2!}x^{n-2}y^2 + \ldots 1\frac{n!}{1!(n-1)!}xy^{n-1} + y^n$$

> *See this Web site for an explanation of the various mathematical symbols used:*
>
> *http://en.wikipedia.org/ wiki/Math_symbols*

Example: $(x + y)^3 = (x + y)(x + y)(x + y)$, and if we multiply, we will find that we have $2^3 = 8$ terms (multiplication of n binomials yields terms):

$$x^3 + yx^2 + yx^2 + yx^2 + y^2x + y^2x + y^2x + y^3 = y^3 + 3yx^2 + 3y^2x + y^3.$$

The binomial theorem tells us how many terms there are of each type.

If we expand a binomial expression of increasing powers, we have a series of polynomials that have a pattern:

$$(x + y)^0 = 1$$
$$(x + y)^1 = 1x + 1y$$
$$(x + y)^2 = 1x^2 + 2xy + 1y^2$$
$$(x + y)^3 = 1x^3 + 3x^2y + 3xy^2 + 1y^3$$
$$(x + y)^4 = 1x^4 + 4x^3y + 4xy^3 + 6x^2y^2 + 1y^4$$
$$(x + y)^5 = 1x^5 + 5x^4y + 5xy^4 + 10x^3y^2 + 10x^2y^3 + 1y^5$$

Example: Expand $(3x + y)^5$.

$$(3x)^5 + \frac{5!}{4!1!}(3x)^4y^1 + \frac{5!}{3!2!}(3x)^3y^2 + \frac{5!}{2!3!}(3x)^2y^3 + \frac{5!}{1!4!}(3x)^1y^4 + y^5 =$$
$$243x^5 + 405x^4y + 270x^3y^2 + 90x^2y^3 + 15xy^4 + y^5$$

Any term of a binomial expansion can be written individually. For example, in the seventh term of $(x + y)^n$ y would be raised to the 6th power, and since the sum of exponents on x and y must equal 7, then the x must be raised to the $n - 6$ power.

The formula to find the rth term of a binomial expansion is

$$\frac{n!}{[n - (r-1)]!(r-1)!}x^{n-(r-1)}y^{r-1}$$

where $r =$ the number of the desired term and $n =$ the power of the binomial.

Example: Find the third term of $(x + 2y)^{11}$.

$x^{n-(r-1)}$	y^{r-1}	Find x and y exponents.
$x^{11-(3-1)}$	y^{3-1}	
x^9	y^2	$y = 2y$
$\frac{11!}{9!2!}(x^9)(2y)^2$		Substitute known values.
$220x^9y^2$		Solution.

Operations in a Complex Field

Adding and subtracting complex numbers

To add or subtract complex numbers, add or subtract the real parts. Then add or subtract the imaginary parts and keep the i (just like combining like terms).

Example: Add (2 + 3i) + (-7 − 4i).

$2 + \text{-}7 = \text{-}5 \qquad\qquad 3i + \text{-}4i = \text{-}i \qquad$ so,

$(2 + 3i) + (\text{-}7 - 4i) = \text{-}5 - i$

Example: Subtract (8 − 5i) − (-3 + 7i)

$8 - 5i + 3 - 7i = 11 - 12i$

Multiplying complex numbers

To multiply two complex numbers, FOIL (F = first terms, O = outer terms, I = inner terms, and L = last terms) the two complex factors together. Replace i^2 with a -1 and finish combining like terms. Answers should have the form $a + b\,$i.

Example: Multiply (8 + 3i)(6 − 2i). Use FOIL (First, Outer, Inner, Last).

Multiply the first terms to get 48.

Multiply the outer terms to get $-16i$ and so on....

$48 - 16i + 18i - 6i^2 \qquad$ Let $i^2 = \text{-}1$.

$48 - 16i + 18i - 6(\text{-}1)$

$48 - 16i + 18i + 6$

$54 + 2i \qquad\qquad$ This is the answer.

Example: Multiply (5 + 8i)² ← Write this out twice.

$(5 + 8i)(5 + 8i) \qquad$ FOIL this.

$25 + 40i + 40i + 64i^2 \qquad$ Let $i^2 = \text{-}1$.

$25 + 40i + 40i + 64(\text{-}1)$

$25 + 40i + 40i - 64$

$\text{-}39 + 80i \qquad\qquad$ This is the answer.

Dividing complex numbers

See Skill 1.16

Solving Polynomial Equations in the Complex Field

Example:

$4x^2 - 2x + 7 = 0$	This is not a perfect square
$x^2 - \frac{1}{2}x + \frac{7}{4} = 0$	Divide by 4, the coefficient of the quadratic term.
$x^2 - \frac{1}{2}x = -\frac{7}{4}$	Put in the form $x^2 + bx$.
$x^2 - \frac{1}{2}x + \frac{1}{16} = -\frac{7}{4} + \frac{1}{16}$	To complete the square, divide b by 2 and then square the result: $\left(-\frac{1}{2} \div 2\right)^2 = \frac{1}{16}$.
$\left(x - \frac{1}{4}\right)^2 = -\frac{27}{16}$	Factor the left side so you have a perfect square; simplify the right.
$\left(x - \frac{1}{4}\right) = \sqrt{-\frac{27}{16}}$	Use the Square Root Property.
$x - \frac{1}{4} = \pm\frac{3i\sqrt{3}}{4}$	Simplify the right side, $\sqrt{-1} = i$.
$x = \frac{1}{4} \pm \frac{3i\sqrt{3}}{4}$	Add $\frac{1}{4}$ to each side to solve for x.

The solution is $x = \{\frac{1}{4} + \frac{3i\sqrt{3}}{4}, \frac{1}{4} - \frac{3i\sqrt{3}}{4}\}$, an imaginary solution with complex numbers.

Check:

Check the solution by graphing, and notice that there are no real solutions for the equation because it has no roots (x-intercepts). Imaginary solutions must be checked by substituting the answers back into the original equation.

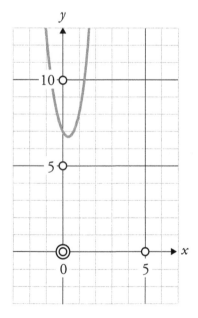

To solve an absolute value equation, follow these steps:

1. Get the absolute value expression alone on one side of the equation.

2. Split the absolute value equation into two separate equations without absolute value bars. In one equation, the expression inside the absolute value bars is equal to the expression on the other side of the original equation. In the other equation, the expression inside the absolute value bars is equal to the negative of the expression on the other side of the original equation.

3. Now solve each of these equations.

4. Check each answer by substituting it into the original equation (with the absolute value symbol). There will be answers that do not check in the original equation. These answers are discarded because they are extraneous solutions. If all of the answers are discarded as incorrect, then the answer to the equation is $\varnothing$, which is the empty or null set. (Answers of 0, 1, or 2 could be correct.)

To solve an absolute value inequality, follow these steps:

1. Get the absolute value expression alone on one side of the inequality. *Remember*: Dividing or multiplying by a negative number will reverse the direction of the inequality sign.

2. Remember what the inequality sign is at this point.

3. Split the absolute value inequality into two separate inequalities. For the first inequality, rewrite the inequality without the absolute value bars and solve it. For the next inequality, write the expression inside the absolute value bars, followed by the opposite inequality sign and then by the negative of the expression on the other side of the inequality. Now solve it.

4. If the inequality sign in step 2 is $<$ or $\leq$, the solution is expressed by connecting the solutions of the two inequalities in step 3 by the word and. The solution set consists of the points between the two numbers on the number line. If the inequality sign in step 2 is $>$ or $\geq$, the solution is expressed by connecting the solutions of the two inequalities in step 3 by the word or. The solution set consists of the points outside the two numbers on the number line.

   If an expression inside an absolute value bar is compared to a negative number, the answer can also be either all real numbers or the empty set ($\varnothing$).

For instance, $|x + 3| < -6$ would have the empty set as the answer, since an absolute value is always positive and will never be less than -6. However, $|x + 3| > -6$ would have all real numbers as the answer, since an absolute value is always positive or at least zero, and will never be less than -6. In similar fashion, $|x + 3| = -6$ would never check because an absolute value will never give a negative value.

Example: Solve and check: $|2x - 5| + 1 = 12.$
$\quad |2x - 5| = 11 \qquad$ Get absolute value alone.

Rewrite as two equations and solve separately:

right-hand side positive		right-hand side negative
$2x - 5 = 11$		$2x - 5 = -11$
$2x = 16$	and	$2x = -6$
$x = 8$		$x = -3$

Check: $|2x - 5| + 1 = 12 \qquad\qquad |2x - 5| + 1 = 12$
$\quad\ |2(8) - 5| + 1 = 12 \qquad\quad |2(-3) - 5| + 1 = 12$
$\qquad\quad |11| + 1 = 12 \qquad\qquad\ |-11| + 1 = 12$
$\qquad\qquad\quad 12 = 12 \qquad\qquad\qquad\quad 12 = 12$

Both 8 and -3 check.

Example: Solve and check: $2|x - 7| - 13 \geq 11.$
$\quad 2|x - 7| \geq 24 \qquad$ Get absolute value alone.
$\quad\ |x - 7| \geq 12$

Rewrite as two inequalities and solve separately:

right-hand side positive		right-hand side negative
$x - 7 \geq 12$	or	$x - 7 \leq -12$
$x \geq 19$	or	$x \leq -5$

Graphing Absolute Value Functions

The absolute value function for a first-degree equation is of the form $y = m|x - h| + k$. Its graph is in the shape of a $\vee$. The point (h, k) is the location of the maximum/minimum point on the graph, and "$\pm m$" are the slopes of the 2 sides of the $\vee$. The graph opens up if m is positive and down if m is negative.

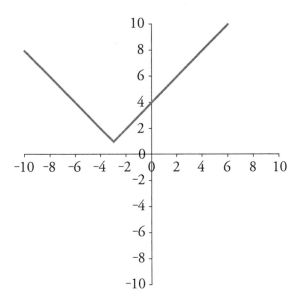

$$y = \left| x + 3 \right| + 1$$

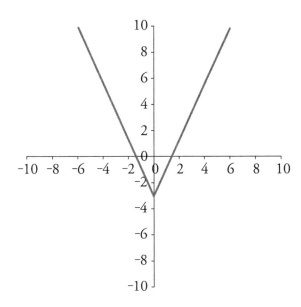

$$y = 2 \left| x \right| - 3$$

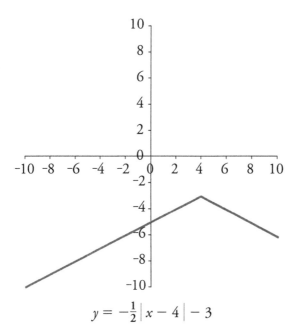

$$y = -\frac{1}{2}\left|x - 4\right| - 3$$

Note that on the first graph, the graph opens up, since m is positive 1. It has (-3,1) as its minimum point. The slopes of the two upward rays are ± 1.

The second graph also opens up, since m is positive. (0,-3) is its minimum point. The slopes of the two upward rays are ± 2.

The third graph is a downward $\wedge$ because m is $-\frac{1}{2}$. The maximum point on the graph is at (4,-3). The slopes of the two downward rays are $\pm\frac{1}{2}$.

The identity function is the linear equation $y = x$. Its graph is a line going through the origin (0,0) and through the first and third quadrants at a 45-degree angle.

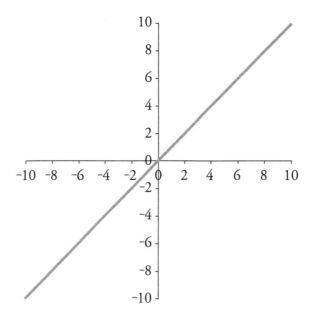

See also Skills 1.20 and 3.4

The general procedure for solving a quadratic equation using a graphing calculator is outlined below. Specific commands are not given since they will be different for different calculators.

1. First express the equation as $ax^2 + bx + c = 0$ if it is not already given in that form.

2. Enter the function in the calculator. e.g. $y = x^2 + x + 1$

3. Press the "Graph" button to display the graph. You can see the points where the graph intersects the x-axis. These are the roots of the equation since $y = 0$ at these values of x.

4. The graph shows you the approximate position of a root. To find the exact value you need to use the "root" function. In many calculators it is CALC-ROOT or CALC-ZERO.

5. When you select the "root" function, you are typically asked to enter upper and lower bounds for the root. You can move a marker on the graph to a point just right of the intersection of the graph and the x-axis and another marker just left of the intersection.

6. The calculator uses the information you have provided to return a root value between the bounds.

7. Repeat the process of setting bounds to identify other roots.

Note:

A. The calculator may not give you an exact value since it uses numerical methods to approximate an answer. You will soon start to recognize that a number such as 0.9999987 is equal to 1.

B. In cases where a quadratic function has complex roots, the graph will not intersect the x-axis. These roots cannot be found using a graphing calculator.

DOMAIN II

GEOMETRY AND MEASUREMENT

PERSONALIZED STUDY PLAN

PAGE	COMPETENCY AND SKILL		KNOWN MATERIAL/ SKIP IT
83	**2:**	**Geometry and measurement**	☐
	2.1:	Solve problems that involve measurement in both metric and traditional systems	☐
	2.2:	Compute perimeter, area, and volume of a variety of figures	☐
	2.3:	Apply the Pythagorean theorem to solve problems	☐
	2.4:	Solve problems involving special triangles	☐
	2.5:	Use relationships such as congruency and similarity to solve problems involving two-dimensional and three-dimensional figures	☐
	2.6:	Solve problems involving parallel and perpendicular lines	☐
	2.7:	Solve problems using the relationships among the parts of triangles	☐
	2.8:	Solve problems using the properties of special quadrilaterals	☐
	2.9:	Solve problems involving angles, diagonals, and vertices of polygons with more than four sides	☐
	2.10:	Solve problems that involve using the properties of circles	☐
	2.11:	Solve problems involving reflections, rotations, and translations of points, lines, or polygons in the plane	☐
	2.12:	Solve problems that can be represented on the XY-plane	☐
	2.13:	Estimate absolute and relative error	☐
	2.14:	Demonstrate an intuitive understanding of a limit	☐
	2.15:	Demonstrate an intuitive understanding of maximum and minimum	☐
	2.16:	Estimate the area of a region in the xy-plane	☐

COMPETENCY 2
GEOMETRY AND MEASUREMENT

SKILL 2.1	Solve problems that involve measurement in both metric and traditional systems

To estimate measurements, you must be familiar with the metric and customary systems.

The following are some common equivalents.

COMMON MEASUREMENT EQUIVALENTS		
ITEM	APPROXIMATELY EQUAL TO	
	Metric	Imperial
Large paper clip	1 gram	0.1 ounce
Average sized adult	75 kilograms	170 pounds
Math textbook	1 kilogram	2 pounds
Thickness of a dime	1 millimeter	0.1 inches
CONVERSION	APPROXIMATELY EQUAL TO	
	Metric	Imperial
Volume	1 quart	1 liter
Distance	1 yard	1 meter
	1 mile	1 kilometer
	1 foot	30 centimeters

Estimate the measurement of the following items:
The length of an adult cow = ___3___ meters
The thickness of a compact disc = ___2___ millimeters
Your height = ___1.5___ meters

The length of your nose = ____4____ centimeters
The weight of your math textbook = ____1____ kilogram
The weight of an automobile = __1,000__ kilogram
The weight of an aspirin = ____1____ gram

Rounding

Given a set of objects and their measurements, using rounding is helpful when you want to find the nearest given unit.

When rounding to a given place value, it is necessary to look at the number in the next smaller place. If this number is 5 or more, the number in the place that is being rounded to is increased by 1, and all of the numbers to the right are changed to zero. If the number is less than 5, the number in the place that is being rounded to stays the same, and all of the numbers to the right are changed to zero.

One method of rounding measurements requires an additional step. First, the measurement must be converted to a decimal number. Then the rules for rounding are applied.

Example: Round the measurements to the given units.

MEASUREMENT	ROUND TO NEAREST	ANSWER
1 foot 7 inches	foot	2 ft
5 pound 6 ounces	pound	5 pounds
$5 \frac{9}{16}$ inches	inch	6 inches

Solution:
Convert each measurement to a decimal number. Then apply the rules for rounding.

1 foot 7 inches $= 1 \frac{7}{12}$ ft $= 1.58333$ ft, round up to 2 ft
5 pounds 6 ounces $= 5 \frac{6}{16}$ pounds $= 5.375$ pound, round to 5 pounds
$5 \frac{9}{16}$ inches $= 5.5625$ inches, round up to 6 inches

Converting Between Measurements

There are many methods for converting measurements in a system. One method is to multiply the given measurement by a conversion factor. This conversion factor is the ratio of

$\frac{\text{new units}}{\text{old units}}$ OR $\frac{\text{what you want}}{\text{what you have}}$

Example: Convert 3 miles to yards.

3 miles × 1,760 yards/mile = ___ yards   1. Multiply by the conversion factor.
= 5,280 yards   2. Cancel the "miles" units.
3. Solve.

Example: Convert 8,750 meters to kilometers.

8,750 meters × 0.001 km/meter = ____ km   1. Multiply by the conversion factor.
= 8.75 km   2. Cancel the "meters" units.
3. Solve.

Approximating Measurements

Most numbers in mathematics are "exact" or "counted," whereas measurements are "approximate." They usually involve interpolation or figuring out which mark on the ruler is closest. Any measurement you get with a measuring device is approximate. Variations in measurement are called precision and accuracy.

PRECISION is a measurement of exactly how a measurement is made, without reference to a true or real value. If a measurement is precise, it can be repeated again and again with little variation in the result. The precision of a measuring device is the smallest fractional or decimal division on the instrument. The smaller the unit or fraction of a unit on the measuring device, the more precisely it can measure.

The greatest possible error of measurement is always equal to one-half the smallest fraction of a unit on the measuring device.

ACCURACY is a measure of how close the result of measurement comes to the "true" value.

If you are throwing darts, the true value is the bull's-eye. If the three darts land on the bull's-eye, the dart thrower is both precise (all land near the same spot) and accurate (the darts all land on the "true" value).

The greatest measure of error allowed is called the TOLERANCE. The least acceptable measure is called the lower limit, and the greatest acceptable measure is called the upper limit. The difference between the upper and lower limits is called the tolerance interval. For example, a specification for an automobile part might be 14.625 ± 0.005 mm. This means that the smallest acceptable length of the part is 14.620 mm and the largest length acceptable is 14.630 mm. The tolerance interval is 0.010 mm. One can see how it would be important for automobile parts to be within a set of limits in terms of length. If the part is too long or too short, it will not fit properly and vibrations may occur, weakening the part and eventually causing damage to other parts.

> **PRECISION:** a measurement of exactly how a measurement is made, without reference to a true or real value

> **ACCURACY:** a measure of how close the result of measurement comes to the "true" value

> **TOLERANCE:** the greatest measure of error allowed

> **SKILL** Compute perimeter and area of triangles, quadrilaterals, circles,
> **2.2** and regions that are combinations of these figures; compute the
> surface area and volume of right prisms, cones, cylinders, spheres
> and solids that are combinations of these figures

Area Problems

Some problems involve computing the area that remains when sections are cut
out of a given figure composed of triangles, squares, rectangles, parallelograms,
trapezoids, or circles. The strategy for solving problems of this nature should be
to identify the given shapes and choose the correct formulas. Subtract the smaller
cut-out shape from the larger shape.

*Example: Find the area of one side of the metal in the circular flat washer
shown below:*

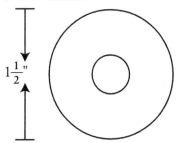

1. The shapes are both circles.

2. Use the formula $A = \pi r^2$ for both.
   (Inside diameter is $\frac{3}{8}$")

Area of larger circle
  $A = \pi r^2$
  $A = \pi(.75^2)$
  $A = 1.77 \text{ in}^2$

Area of smaller circle
  $A = \pi r^2$
  $A = \pi(.1875^2)$
  $A = .11 \text{ in}^2$

Area of metal washer = larger area − smaller area
  $= 1.77 \text{ in}^2 - .11 \text{ in}^2$
  $= 1.66 \text{ in}^2$

*Example: You have decided to fertilize your lawn. The shapes and dimensions
of your lot, house, pool, and garden are given in the diagram below. The
shaded area will not be fertilized. If each bag of fertilizer costs $7.95 and
covers 4,500 square feet, find the total number of bags needed and the total
cost of the fertilizer.*

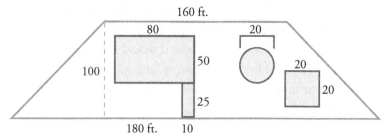

Area of Lot	Area of House	Area of Driveway	Area of Garden	Area of Pool
$A = \frac{1}{2}h(b_1 + b_2)$	$A = LW$	$A = LW$	$A = s^2$	$A = \pi r^2$
$A = \frac{1}{2}(100)(180 + 160)$	$A = (80)(50)$	$A = (10)(25)$	$A = (20)^2$	$A = \pi(10)^2$
$A = 17,000$ sq ft	$A = 4,000$ sq ft	$A = 250$ sq ft	$A = 400$ sq ft	$A = 314.159$ sq ft

Total area to fertilize = Lot area − (House + Driveway + Pool + Garden)
= 17,000 − (4,000 + 250 + 314.159 + 400)
= 12,035.841 sq ft

Number of bags needed = Total area to fertilize/4,500 sq ft. bag
= 12,035.841/4,500
= 2.67 bags

Since we cannot purchase 2.67 bags we must purchase 3 full bags.

Total cost = Number of bags × $7.95
= 3 × $7.95
= $23.85

Examining the change in area or volume of a given figure requires first finding the existing area given the original dimensions and then finding the new area given the increased dimensions.

Example: Given the rectangle below, determine the change in area if the length is increased by 5 and the width is increased by 7.

7

4

Draw and label a sketch of the new rectangle.

12

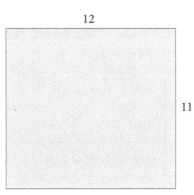

11

Find the areas.

Area of original = *LW*
  = (7)(4)
  = 28 units²

Area of enlarged shape = *LW*
  = (12)(11)
  = 132 units²

The change in area is 132 − 28 = 104 units².

To find the area of a compound shape, cut the compound shape into smaller, more familiar shapes, and then compute the total area by adding the areas of the smaller parts.

Example: Find the area of the given shape.

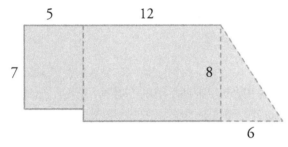

1. Using a dotted line, we have cut the shape into smaller parts that are familiar

2. Use the appropriate formula for each shape and find the sum of all areas

Area 1 = *LW*
  = (5)(7)
  = 35 units²

Area 2 = *LW*
  = (12)(8)
  = 96 units²

Area 3 = $\frac{1}{2}bh$
  = $\frac{1}{2}$(6)(8)
  = 24 units²

Total area = Area 1 + Area 2 + Area 3
  = 35 + 96 + 24
  = 155 units²

Volume and Surface Area Problems

Use the following formulas to find volume and surface area.

FIGURE	VOLUME	TOTAL SURFACE AREA
Right Cylinder	$\pi r^2 h$	$2\pi r h + 2\pi r^2$
Right Cone	$\frac{\pi r^2 h}{3}$	$\pi r \sqrt{r^2 + h^2} + \pi r^2$
Sphere	$\frac{4}{3}\pi r^3$	$4\pi r^2$
Rectangular Solid	LWH	$2LW + 2WH + 2LH$

Note: $\sqrt{r^2 + h^2}$ is equal to the slant height of the cone.

Example: Given the figure below, find the volume and the surface area.

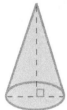

$r = 5$ in
$h = 6.2$ in

Volume $= \frac{\pi r^2 h}{3}$ First write the formula.
$= \frac{1}{3}\pi(5^2)(6.2)$ Then substitute.
$= 162.23333$ cubic inches   Compute.

Surface area $= \pi r \sqrt{r^2 + h^2} + \pi r^2$ First write the formula.
$= \pi 5\sqrt{5^2 + 6.2^2} + \pi 5^2$ Then substitute.
$= 203.549$ square inches   Compute.

> **Note:** *volume is always given in cubic units, and area is always given in square units.*

Example: A water company is trying to decide whether to use traditional cylindrical paper cups or to offer conical paper cups; they both cost the same. The traditional cups are 8 cm wide and 14 cm high. The conical cups are 12 cm wide and 19 cm high. The company will use the cup that holds more water.

Draw and label a sketch of each cup.

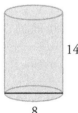

$V = \pi r^2 h$	$V = \frac{\pi r^2 h}{3}$	1. Write a formula.
$V = \pi(4)^2(14)$	$V = \frac{1}{3}\pi(6)^2(19)$	2. Substitute.
$V = 703.36$ cm^3	$V = 715.92$ cm^3	3. Solve.

The choice should be the conical cup, since its volume is greater.

Example: How much material is needed to make a basketball that has a diameter of 15 inches? How much air is needed to fill the basketball?

Draw and label a sketch:

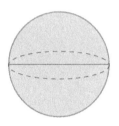

$D = 15$ inches

Total surface area

$TSA = 4\pi r^2$

$= 4\pi(7.5)^2$

$= 706.5 \text{ in}^2$

Volume

$V = \frac{4}{3}\pi r^3$

$= \frac{4}{3}\pi(7.5)^3$

$= 1766.25 \text{ in}^3$

1. Write a formula.

2. Substitute.

3. Solve.

Area and Perimeter Problems

FIGURE	AREA FORMULA	PERIMETER FORMULA
Rectangle	LW	$2(L + W)$
Triangle	$\frac{1}{2}bh$	sum of lengths of sides
Parallelogram	bh	sum of lengths of sides
Trapezoid	$\frac{1}{2}h(b_1 + b_2)$	sum of lengths of sides

Example: Find the area and perimeter of a rectangle if its length is 12 inches and its diagonal is 15 inches.

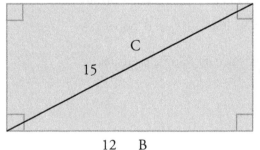

1. Draw and label sketch.
2. Since the height is still needed, use the Pythagorean formula to find missing leg of the triangle.

$A^2 + B^2 = C^2$

$A^2 + 12^2 = 15^2$

$A^2 = 15^2 - 12^2$

$A^2 = 81$

$A = 9$

Now use this information to find the area and perimeter.

A = *LW*	P = 2(*L* + *W*)	1. Write formula.
A = (12)(9)	P = 2(12 + 9)	2. Substitute.
A = 108 in²	P = 42 inches	3. Solve.

Circles

Given a circular figure, the formulas are as follows:

$$A = \pi r^2 \quad C = \pi d \ \text{or} \ 2\pi r$$

Example:

If the area of a circle is 50 cm², find the circumference.

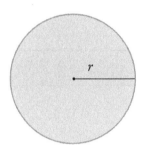

A = 50 cm²

1. Draw a sketch.

2. Determine what is still needed.

Use the area formula to find the radius.

$A = \pi r^2$	1. Write the formula.
$50 = \pi r^2$	2. Substitute.
$\frac{50}{\pi} = r^2$	3. Divide by π.
$15.924 = r^2$	4. Substitute.
$\sqrt{15.924} = \sqrt{r^2}$	5. Take the square root of both sides.
$3.99 \approx r$	6. Compute.

Use the approximate answer (due to rounding) to find the circumference.

$A = 2\pi r$	1. Write the formula.
$C = 2\pi(3.99)$	2. Substitute.
$C \approx 25.057$	3. Compute.

When using formulas to solve a geometry problem, it is helpful to use the same strategies used for general problem solving. First, draw and label a sketch if needed. Second, write down the formula and then substitute in the known values. This will assist in identifying what is still needed (the unknown). Finally, solve the resulting equation.

Being consistent in the strategic approach to problem solving is paramount to teaching the concept as well as solving it.

Example: Use the appropriate problem-solving strategies to find the solution.

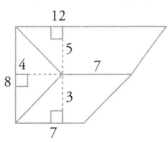

1. Find the area of the given figure.
2. Cut the figure into familiar shapes.
3. Identify what types of figures are given and write the appropriate formulas.
   All dimensions are given in feet.

AREA OF FIGURE 1 (TRIANGLE)	AREA OF FIGURE 2 (PARALLELOGRAM)	AREA OF FIGURE 3 (TRAPEZOID)
$A = \frac{1}{2}bh$	$A = bh$	$A = \frac{1}{2}h(a + b)$
$A = \frac{1}{2}(8)(4)$	$A = (7)(3)$	$A = \frac{1}{2}(5)(12 + 7)$
$A = 16$ sq. ft	$A = 21$ sq. ft	$A = 47.5$ sq. ft

Now find the total area by adding the area of all figures:

Total area $= 16 + 21 + 47.5$
$= 84.5$ sq. ft

Example: Given the figure below, find the area by dividing the polygon into smaller shapes.

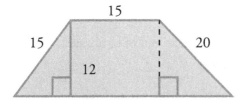

1. Divide the figure into two triangles and a rectangle.
2. Find the missing lengths.
3. Find the area of each part.
4. Find the sum of all of the areas.

Find the base of both right triangles using the Pythagorean formula:

$$a^2 + b^2 = c^2 \qquad\qquad a^2 + b^2 = c^2$$
$$a^2 + 12^2 = 15^2 \qquad\quad a^2 + 12^2 = 20^2$$
$$a^2 = 225 - 144 \qquad\quad a^2 = 400 - 144$$
$$a^2 = 81 \qquad\qquad\quad a^2 = 256$$
$$a = 9 \qquad\qquad\qquad a = 16$$

AREA OF TRIANGLE 1	AREA OF TRIANGLE 2	AREA OF RECTANGLE
$A = \frac{1}{2}bh$	$A = \frac{1}{2}bh$	$A = LW$
$A = \frac{1}{2}(9)(12)$	$A = \frac{1}{2}(16)(12)$	$A = (15)(12)$

A = 54 sq. units	A = 96 sq. units	A = 180 sq. units

Find the sum of all three figures:

54 + 96 + 180 = 330 square units

Surface Area and Volume of Prisms and Pyramids

To compute the surface area and volume of right prisms, cones, cylinders, spheres, and solids that are combinations of these figures, use the following formulas:

FIGURE	LATERAL AREA	TOTAL AREA	VOLUME
Right Prism	Ph	2B + Ph	Bh
Regular Pyramid	$\frac{1}{2}Pl$	$\frac{1}{2}Pl + B$	$\frac{1}{3}Bh$

P = Perimeter; h = height; B = Area of Base; l = slant height

Example: Find the total area of the given figure.

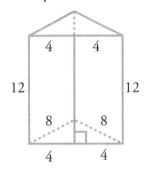

1. Since this is a triangular prism, first find the area of the bases.
2. Find the area of each rectangular lateral face.
3. Add the areas together.

$A = \frac{1}{2}bh$ $A = LW$ 1. Write the formula.

$8^2 = 4^2 + h^2$ 2. Find the height of the base triangle.

                       $A = (8)(12)$ 3. Substitute the known values.

$A = 27.713$ sq. units $A = 96$ sq. units   4. Compute.

Total Area = 2(27.713) + 3(96)

            = 343.426 sq. units

The Pythagorean Theorem

> **PYTHAGOREAN THEOREM:** states that in a right triangle, the square of the length of the hypotenuse is equal to the sum of the squares of the lengths of the legs

The PYTHAGOREAN THEOREM states that in a right triangle, the square of the length of the hypotenuse is equal to the sum of the squares of the lengths of the legs. Symbolically, this is stated as follows:

$$c^2 = a^2 + b^2$$

Example: Given the right triangle below, find the missing side.

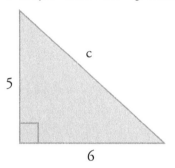

$c^2 = a^2 + b^2$	1. Write the formula.
$c^2 = 5^2 + 6^2$	2. Substitute the known values.
$c^2 = 61$	3. Take the square root.
$c^2 = \sqrt{61}$ or 7.81	4. Solve.

Converse of the Pythagorean Theorem

The converse of the Pythagorean Theorem states that if the square of one side of a triangle is equal to the sum of the squares of the other two sides, then the triangle is a right triangle.

Example: Given △XYZ with sides measuring 12, 16, and 20 cm, is this a right triangle?

$$c^2 = a^2 + b^2$$
$$20^2 \ \underline{?} \ 12^2 + 16^2$$
$$400 \ \underline{?} \ 144 + 256$$
$$400 = 400$$

Yes, the triangle is a right triangle.

This theorem can be expanded to determine if triangles are obtuse or acute:

If the square of the longest side of a triangle is greater than the sum of the squares of the other two sides, then the triangle is an obtuse triangle.

and

If the square of the longest side of a triangle is less than the sum of the squares of the other two sides, then the triangle is an acute triangle.

Example: Given △LMN with sides measuring 7, 12, and 14 inches, is the triangle right, acute, or obtuse?

$$14^2 \underline{\ ?\ } 7^2 + 12^2$$
$$196 \underline{\ ?\ } 49 + 144$$
$$196 > 193$$

Therefore, the triangle is obtuse.

SKILL 2.4 **Solve problems involving special triangles, such as isosceles and equilateral**

Classifying Triangles

A TRIANGLE is a polygon with three sides. Triangles can be classified by the types of angles or the lengths of their sides.

Classifying by angles

An ACUTE TRIANGLE has exactly three *acute* angles.

A RIGHT TRIANGLE has one *right* angle.

An OBTUSE TRIANGLE has one *obtuse* angle.

  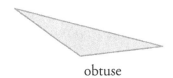

acute   right   obtuse

Classifying by sides

All *three* sides of an EQUILATERAL TRIANGLE are the same length.

Two sides of an ISOSCELES TRIANGLE are the same length.

TRIANGLE: a polygon with three sides

ACUTE TRIANGLE: a triangle with exactly three acute angles

RIGHT TRIANGLE: a triangle with one right angle

OBTUSE TRIANGLE: a triangle with one obtuse angle

EQUILATERAL TRIANGLE: a triangle with all three sides the same length

ISOSCELES TRIANGLE: a triangle with two sides the same length

SCALENE TRIANGLE: a triangle with no sides the same length

The sum of the measures of the angles of a triangle is 180°.

None of the sides of a SCALENE TRIANGLE are the same length.

  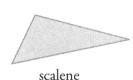

equilateral   isosceles   scalene

Properties of Triangles

The sum of the measures of the angles of a triangle is 180°.

Problems involving a special triangle can be solved by taking advantage of known properties of these triangles. For instance, the angles (in addition to the sides) of an equilateral triangle are all congruent, and the two angles formed by the two congruent sides and the noncongruent side of an isosceles triangle are congruent. These and other properties of special triangles are often the key to solving such problems.

Example: Find the angles α and β of the following triangle.

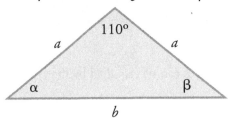

Because the triangle has two equal sides, it is isosceles. Consequently, angles α and β are congruent.

$\alpha + \beta + 110° = 180°$ The sum of the angles of a triangle is 180°.

$\alpha + \beta = 70°$

$\alpha = \beta = 35°$ The triangle above is isosceles; solve for either α or β.

Example: Find the length x of the altitude of an equilateral triangle with sides of length a.

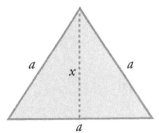

The line segment that defines the altitude of an equilateral triangle bisects the triangle. As a result, two congruent right triangles are formed, each having a hypotenuse a and a shorter leg $\frac{a}{2}$. Use the Pythagorean Theorem to find x.

$$x^2 = a^2 - \left(\frac{a}{2}\right)^2$$

$$x^2 = a^2 - \frac{a^2}{4} = \frac{3a^2}{4}$$

$$x = \frac{\sqrt{3}a}{2}$$

The answer to the problem is thus determined.

Other problems involving special triangles follow a similar pattern: first, identify the unique properties of the triangle, and then apply them to the facts of the problem.

SKILL 2.5 Use relationships such as congruency and similarity to solve problems involving two-dimensional and three-dimensional figures

Congruency

CONGRUENT figures have the same size and shape. If one is placed above the other, it will fit exactly. Congruent lines have the same length. Congruent angles have equal measures.

The symbol for congruent is $\cong$.

Polygons (pentagons) *ABCDE* and *VWXYZ* are congruent. They are exactly the same size and shape.

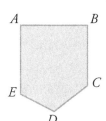 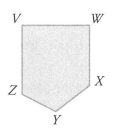

$$ABCDE \cong VWXYZ$$

Corresponding parts are those congruent angles and congruent sides, as follows:

Corresponding angles	Corresponding sides
$\angle A \leftrightarrow \angle V$	$AB \leftrightarrow VW$
$\angle B \leftrightarrow \angle W$	$BC \leftrightarrow WX$
$\angle C \leftrightarrow \angle X$	$CD \leftrightarrow XY$
$\angle D \leftrightarrow \angle Y$	$DE \leftrightarrow YZ$
$\angle E \leftrightarrow \angle Z$	$AE \leftrightarrow VZ$

Similarity

Two figures that have the same shape but not necessarily the same size are SIMILAR. Two polygons are similar if corresponding angles are congruent and corresponding sides are in proportion. Corresponding parts of similar polygons are proportional. For example, the following two quadrilaterals are similar.

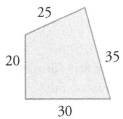 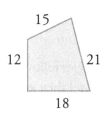

These relationships among figures can apply to three-dimensional figures as well as to two-dimensional figures. For instance, two parallelepipeds can be similar if the corresponding angles are congruent and the corresponding sides are all proportional, as shown below.

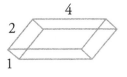 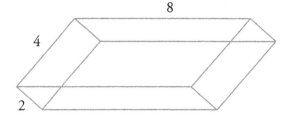

Using Congruence and Similarity to Solve Problems

By using the congruence and similarity properties of both two- and three-dimensional figures, problems involving such figures can often be solved more easily.

Example: The triangles shown below each have base angles equal in measure to each other and to the base angles of the other triangle. Find the area of the larger triangle. The altitude of the smaller triangle is shown as a dashed line and has length 3.

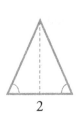 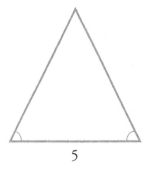

Since both of the triangles have congruent base angles, the two triangles are both isosceles. Furthermore, the apex angles are also congruent. As a result, the two triangles are similar. Calculate the corresponding altitude x of the larger triangle by using the proportion determined by the bases.

$$\frac{5}{2} = \frac{\text{alt}_{\text{larger}}}{\text{alt}_{\text{smaller}}} = \frac{x}{3}$$

$$x = 7.5$$

Finally, calculate the area of the larger triangle.

$$A = \frac{1}{2}bh = \frac{1}{2}(5)(7.5) = 18.75$$

Thus, the answer is 18.75 square units.

The same type of approach can be used to solve problems with congruent or similar figures in two or three dimensions.

SKILL 2.6 Solve problems involving parallel and perpendicular lines

Parallel and Perpendicular Lines

Parallel lines do not intersect.

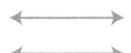

Perpendicular lines form a 90-degree angle to each other.

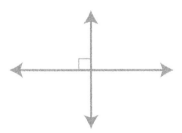

A number of properties of parallel lines (when intersected by a nonparallel line) can be used when solving problems. These properties include the following.

Properties of Parallel Lines

Parallel Lines Postulate: If two lines are parallel and are cut by a transversal, corresponding angles have the same measure. Corresponding angles are in the corresponding positions on two parallel lines cut by a transversal, as shown below.

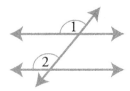

Alternate Interior Angles Theorem: If two parallel lines are cut by a transversal, the alternate interior angles are congruent.

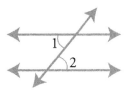

Alternate exterior angles are diagonal angles on the outside of two lines cut by a transversal.

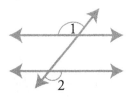

These relationships among parallel and transversal lines can be of immense help in solving some geometry problems. Consider the following examples.

Example: Prove that the sum of the interior angles of a triangle is always 180°.
Consider a triangle with arbitrary angles α, β, and γ. Draw two parallel lines such that one parallel line coincides with any side of the triangle and the other parallel line intersects the vertex opposite that side.

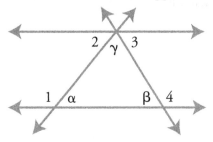

Proof: $\angle\alpha = \angle 2$ Alternate interior angles.
      $\angle\beta = \angle 3$ Alternate interior angles.
      $\angle 2 + \angle 3 + \angle\gamma = 180°$ Supplementary angles
                            (a line sweeps out 180°).
      $\angle\alpha + \angle\beta + \angle\gamma = 180°$ Substitution.

Thus, the sum of all the interior angles of a triangle is always 180 degrees.

Example: Prove that the two triangles formed by the two parallel lines, each cut by two transversals, are similar.

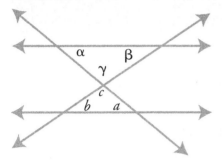

To prove that the triangles are similar, it is sufficient to show that the corresponding angles are congruent.

$\angle c \cong \angle \gamma$ Vertical angles (opposite angles formed by the intersection of two lines) are congruent.

$\angle a = \angle \alpha$ Alternate interior angles.

$\angle b = \angle \beta$ Alternate interior angles.

Thus, since the corresponding angles of the two triangles are congruent, the triangles are similar.

Properties of Perpendicular Lines

The characteristics of perpendicular lines can also be used in solving problems.

Example: Find angle α given that angle β is 56°.

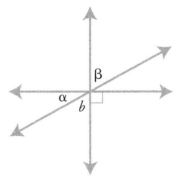

Angle α can be found by first noting that angles b and β are congruent (vertical angles). The sum of angles α and b must be 90°, since the two lines that define this sum of angles are perpendicular. Then perform the simple calculation for α:

$\alpha = 90° - b = 90° - \beta = 90° - 56° = 34°$

The result is then 34°.

> SKILL **Solve problems using the relationships among the parts of**
> 2.7 **triangles, such as sides, angles, medians, midpoints, and**
> **altitudes**

The Parts of Triangles

The segment joining the midpoints of two sides of a triangle is parallel to the third side of the triangle and is one-half the length of the third side of the triangle.

Every angle has exactly one ray that bisects the angle. If a point on such a bisector is located, then the point is equidistant from the two sides of the angle. Distance from a point to a side is measured along a segment that is perpendicular to the angle's side. The converse is also true: If a point is equidistant from the sides of an angle, then the point is on the bisector of the angle.

Every segment has exactly one line that is both perpendicular to and bisects the segment. If a point on such a perpendicular bisector is located, then the point is equidistant to the endpoints of the segment. The converse is also true: If a point is equidistant from the endpoints of a segment, then that point is on the perpendicular bisector of the segment.

If the three segments that bisect the three angles of a triangle are drawn, the segments will all intersect in a single point. This point is equidistant from all three sides of the triangle. Recall that the distance from a point to a side is measured along the perpendicular from the point to the side.

An ALTITUDE of a triangle is a segment that extends from one vertex and is perpendicular to the side opposite that vertex. In some cases, the side opposite the vertex used will need to be extended in order for the altitude to form a perpendicular to the opposite side. The length of the altitude is used when referring to the height of the triangle.

If three or more segments intersect in a single point, the point is called a POINT OF CONCURRENCY.

The following sets of special segments all intersect in points of concurrency:

- Angle bisectors
- Medians
- Altitudes
- Perpendicular bisectors

ALTITUDE: a segment that extends from one vertex and is perpendicular to the side opposite that vertex

POINT OF CONCURRENCY: the point at which three or more segments intersect

The points of concurrency can lie inside the triangle, outside the triangle, or on one of the sides of the triangle. The following table summarizes this information.

POSSIBLE LOCATIONS OF THE POINTS OF CONCURRENCY			
	Inside the Triangle	Outside the Triangle	On the Triangle
Angle Bisectors	X		
Medians	X		
Altitudes	X	X	X
Perpendicular Bisectors	X	X	X

Recall that a MEDIAN is a segment that connects a vertex to the midpoint of the side opposite that vertex. Every triangle has three medians. The point of concurrency of the three medians of a triangle is called the CENTROID.

The centroid divides each median into two segments whose lengths are always in the ratio of 1:2. The distance from the vertex to the centroid is always twice the distance from the centroid to the midpoint of the side opposite the vertex.

MEDIAN: a segment that connects a vertex to the midpoint of the side opposite that vertex

CENTROID: the point of concurrency of the three medians of a triangle

Congruent Triangles

Two triangles are congruent if each of the three angles and three sides of one triangle match up in a one-to-one fashion with corresponding angles and sides of the second triangle. In order to see how the sides and angles match up, it is sometimes necessary to imagine rotating or reflecting one of the triangles so the two figures are oriented in the same position.

Some problems require proving that two triangles are congruent or using congruence to answer a question. The following postulates and theorems can be used in problems involving congruent triangles.

The SSS postulate

If three sides of one triangle are congruent to three sides of another triangle, then the two triangles are congruent.

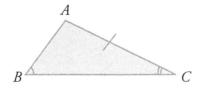
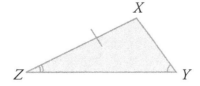

Since $AB \cong XY$, $BC \cong YZ$, and $AC \cong XZ$, then $\triangle ABC \cong \triangle XYZ$.

Example: Given isosceles triangle ABC with D the midpoint of base AC, prove the two triangles formed by AD are congruent.

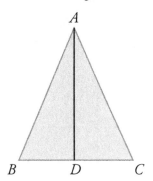

Proof:

1.	Isosceles triangle *ABC*, *D* midpoint of base *AC*	Given.
2.	$AB \cong BC$	An isosceles $\triangle$ has two congruent sides.
3.	$BD \cong DC$	Midpoint divides a line into two equal parts.
4.	$AD \cong AD$	Reflexive.
5.	$\triangle ABD \cong \triangle ACD$	SSS.

The SAS postulate

If two sides and the included angle of one triangle are congruent to two sides and the included angle of another triangle, then the two triangles are congruent.

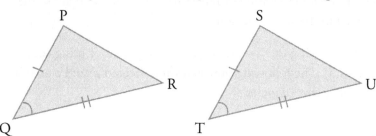

Example:

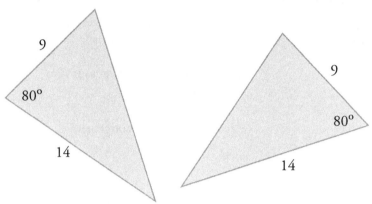

The two triangles are congruent by SAS.

The ASA postulate

If two angles and the included side of one triangle are congruent to two angles and the included side of another triangle, the triangles are congruent.

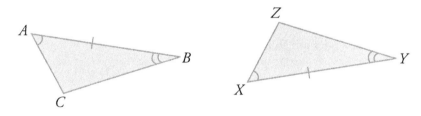

$\angle A \cong \angle X$, $\angle B \cong \angle Y$, $AB \cong XY$, then $\triangle ABC \cong \triangle XYZ$ by ASA

Example: Given two right triangles with one leg of each measuring 6 cm and the adjacent angle 37°, prove that the triangles are congruent.

1. Right triangles *ABC* and *KLM*   Given.

   $AB = KL = 6$ cm
   $\angle A = \angle K = 37°$

2. $AB \cong KL$ Figures with the same measure are
                     congruent.
   $\angle A \cong \angle K$

3. $\angle B \cong \angle L$ All right angles are congruent.

4. $\triangle ABC \cong \triangle KLM$ ASA.

Example: What method would you use to prove that the triangles are congruent?

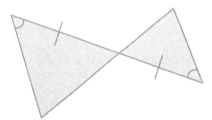

ASA because the vertical angles are congruent.

The AAS theorem

If two angles and a nonincluded side of one triangle are congruent to the corresponding parts of another triangle, then the triangles are congruent.

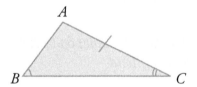 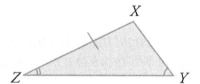

$\angle B \cong \angle Y$, $\angle C \cong \angle Z$, $AC \cong XZ$, then $\triangle ABC \cong \triangle XYZ$ by AAS.

We can derive this theorem because if two angles of the triangles are congruent, then the third angle must also be congruent. Therefore, we can use the ASA postulate.

The HL theorem

If the hypotenuse and a leg of one right triangle are congruent to the corresponding parts of another right triangle, the triangles are congruent.

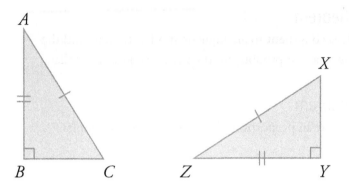

Since $\angle B$ and $\angle Y$ are right angles and $AC \cong XZ$ (hypotenuse of each triangle), $AB \cong YZ$ (corresponding leg of each triangle), then $\triangle ABC \cong \triangle XYZ$ by the HL theorem.

Example: What method would you use to prove that the triangles are congruent?

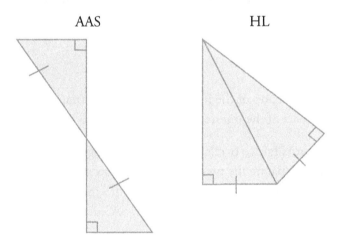

Similar Triangles

For a discussion of similarity see Skill 2.5. The following postulates and theorems can be used to show the similarity of two triangles.

The AA similarity postulate

If two angles of one triangle are congruent to two angles of another triangle, then the triangles are similar.

The SAS similarity theorem

If an angle of one triangle is congruent to an angle of another triangle and the sides adjacent to those angles are in proportion, then the triangles are similar.

The SSS similarity theorem

If the sides of two triangles are in proportion, then the triangles are similar.

Example:

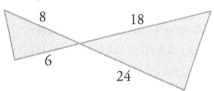

Figure is not drawn to scale.

The two triangles are similar by the SAS similarity theorem, since the sides are proportional and the vertical angles are congruent.

Overlapping Triangles

RIGHT TRIANGLE: a triangle with one right angle

Two triangles are overlapping if a portion of the interior region of one triangle is shared in common with all or a part of the interior region of the second triangle.

HYPOTENUSE: the side opposite the right angle

The most effective method for proving two overlapping triangles congruent is to draw the two triangles separated. Separate the two triangles and label all of the vertices using the labels from the original overlapping figures. Once the separation is complete, apply one of the congruence shortcuts: SSS, ASA, SAS, AAS, or HL.

LEGS: the two sides of a right triangle that are adjacent to the right angle

Right Triangles

A RIGHT TRIANGLE is a triangle with one right angle. The side opposite the right angle is called the HYPOTENUSE. The other two sides are the LEGS. An ALTITUDE is a line drawn from one vertex, perpendicular to the opposite side.

ALTITUDE: a line drawn from one vertex, perpendicular to the opposite side

When an altitude is drawn to the hypotenuse of a right triangle, then the two triangles formed are similar to the original triangle and to each other.

Example:

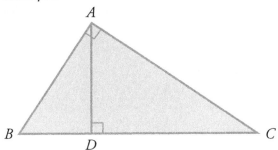

Given right triangle *ABC* with right angle at *A*, altitude *AD* drawn to hypotenuse *BC* at *D*.

$\triangle ABC \sim \triangle ABD \sim \triangle ACD$. The triangles formed are similar to each other and to the original right triangle.

If *a*, *b*, and *c* are positive numbers such that $\frac{a}{b} = \frac{b}{c}$, then *b* is called the GEOMETRIC MEAN between *a* and *c*.

> **GEOMETRIC MEAN:** a measure of central tendency

Example: Find the geometric mean between 6 and 30.

$$\frac{6}{x} = \frac{x}{30}$$
$$x^2 = 180$$
$$x = \sqrt{180} = \sqrt{36 \cdot 5} = 6\sqrt{5}$$

The geometric mean is significant when the altitude is drawn to the hypotenuse of a right triangle:

The length of the altitude is the geometric mean between each segment of the hypotenuse.

AND

Each leg is the geometric mean between the hypotenuse and the segment of the hypotenuse that is adjacent to the leg.

Example:

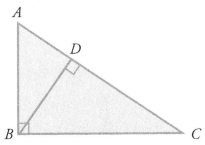

Given:

$\triangle ABC$ is a right triangle.

BD is the altitude of $\triangle ABC$.

$AB = 6$

$AC = 12$

Find *AD, CD, BD,* and *BC.*

$$\frac{12}{6} = \frac{6}{AD} \qquad \frac{3}{BD} = \frac{BD}{9} \qquad \frac{12}{BC} = \frac{BC}{9}$$

$$12(AD) = 36 \qquad (BD)^2 = 27 \qquad (BC)^2 = 108$$

$AD = 3$

$BD = \sqrt{27} = \sqrt{9 \cdot 3} = 3\sqrt{3}$

$BC = \sqrt{108} = \sqrt{36 \cdot 3} = 6\sqrt{3}$

$CD = 12 - 3 = 9$

Special right triangles

Given the special right triangles below, we can find the lengths of other special right triangles.

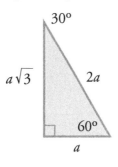 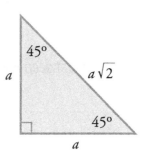

Example:

If $8 = a\sqrt{2}$, then $a = \frac{8}{\sqrt{2}}$ or 5.657.

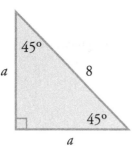

If $7 = a$, then $c = a\sqrt{2} = 7\sqrt{2}$ or 9.899.

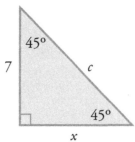

If $2a = 10$, then $a = 5$ and $x = a\sqrt{3} = 5\sqrt{3}$ or 8.66.

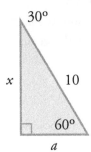

Trigonometric Relationships

Given right triangle *ABC*, the adjacent side and opposite side can be identified for each angle *A* and *B*.

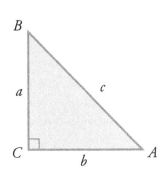

Looking at angle *A*, it can be determined that side *b* is adjacent to angle *A* and side *a* is opposite angle *A*.

If we now look at angle *B*, we see that side *a* is adjacent to angle *B* and side *b* is opposite angle *B*.

The longest side (opposite the 90-degree angle) is always called the hypotenuse.

The basic trigonometric ratios are as follows:

$$\text{Sine} = \frac{\text{opposite}}{\text{hypotenuse}} \qquad \text{Cosine} = \frac{\text{adjacent}}{\text{hypotenuse}} \qquad \text{Tangent} = \frac{\text{opposite}}{\text{adjacent}}$$

Example: Use triangle ABC to find the sin, cos, and tan for angle A.

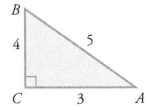

$$\sin A = \frac{4}{5}$$
$$\cos A = \frac{3}{5}$$
$$\tan A = \frac{4}{3}$$

Use the basic trigonometric ratios of sine, cosine, and tangent to solve for the missing sides of right triangles when given at least one of the acute angles. The sine, cosine, and tangent of any angle can be found using a scientific calculator.

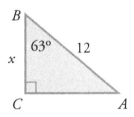

In the triangle *ABC*, an acute angle is 63 degrees, and the length of the hypotenuse is 12. The missing side is the one adjacent to the given angle.

The appropriate trigonometric ratio to use would be cosine, since we are looking for the adjacent side and we have the length of the hypotenuse.

$$\text{Cos B} = \frac{\text{adjacent}}{\text{hypotenuse}} \qquad \text{1. Write the formula.}$$

$$\text{Cos } 63 = \frac{x}{12} \qquad \text{2. Substitute the known values.}$$

$$0.454 = \frac{x}{12}$$

$$x = 5.448 \qquad \text{3. Solve.}$$

Example: Find the missing angle.

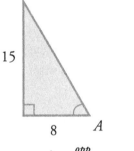

$$\tan A = \frac{opp}{adj} \qquad \tan A = \frac{15}{8} = 1.875$$

Looking on the trigonometric chart or using the inverse tan function on a scientific calculator, the angle whose tangent is closest to 1.875 is 62°. Thus, $\angle A \approx 62°$.

Example: Find the missing side.

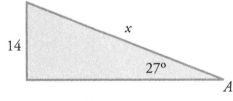

$$\sin A = \frac{opp}{hyp}$$

$$\sin 27° = \frac{14}{x}$$

$$0.4540 \approx \frac{14}{x}$$

$$x \approx \frac{14}{.454}$$

$$x \approx 30.8$$

Example: Find the missing side.

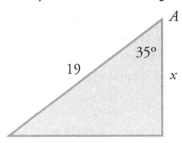

$\cos A = \frac{add}{hyp}$

$\cos 35° = \frac{x}{19}$

$x \approx 19 \times 0.82$

$x \approx 15.6$

SKILL 2.8 **Solve problems using the properties of special quadrilaterals, such as the square, rectangle, parallelogram, rhombus, and trapezoid; describe relationships among sets of special quadrilaterals**

A QUADRILATERAL is a polygon with four sides. The sum of the measures of the angles of a quadrilateral is 360°. Although quadrilaterals encompass such figures as squares and rectangles, they also include many other special types of four-sided polygons. The following two shapes are both quadrilaterals, although they have no apparent special characteristics.

QUADRILATERAL: a polygon with four sides

TEACHER CERTIFICATION STUDY GUIDE

113

Relationships Among Special Quadrilaterals

The different types of special quadrilaterals include several well-known figures. Some quadrilaterals are included in several different sets. For example, a square is also a rectangle and a rhombus as well as a parallelogram, and it has all of the characteristics of these different figures. The following set diagram shows the relationships among special quadrilaterals.

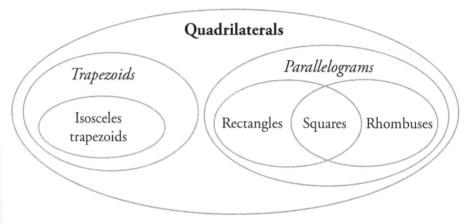

TRAPEZOID: a quadrilateral with exactly one pair of parallel sides (called bases)

A TRAPEZOID (shown below) is a quadrilateral with exactly one pair of parallel sides (called bases). The nonparallel sides are called legs.

ALTITUDE: a line segment drawn from a point on either base, perpendicular to the opposite base

MEDIAN: a line segment that joins the midpoints of each leg

An ALTITUDE is a line segment drawn from a point on either base, perpendicular to the opposite base. The MEDIAN is a line segment that joins the midpoints of each leg. The median of a trapezoid is parallel to the bases and equal to one-half the sum of the lengths of the bases.

In an ISOSCELES TRAPEZOID, the nonparallel sides (legs) are congruent.

ISOSCELES TRAPEZOID: a quadrilateral where the nonparallel sides are congruent

Isosceles trapezoids exhibit all of the properties of ordinary trapezoids, plus the following special properties:

- The base angles are congruent
- The diagonals are congruent

A PARALLELOGRAM is a quadrilateral with two pairs of parallel sides.

Parallelograms exhibit the following characteristics:

- The diagonals bisect each other
- Each diagonal divides the parallelogram into two congruent triangles
- Both pairs of opposite sides are congruent
- Both pairs of opposite angles are congruent
- Any two adjacent angles are supplementary

A RHOMBUS is a parallelogram with all sides of equal length.

Rhombuses have all of the properties of parallelograms.

In addition, the following properties also apply:

- All sides are congruent
- The diagonals are perpendicular
- The diagonals bisect the angles

A RECTANGLE is a parallelogram in which each interior angle is 90°.

Rectangles have all of the properties of parallelograms.

In addition, rectangles have the following properties:

- All interior angles are right angles
- The diagonals are congruent

A SQUARE is a rectangle with all sides of equal length.

Squares have all of the properties of rectangles and all of the properties of rhombuses.

The following table summarizes the properties of each type of parallelogram.

	Parallel Opposite Sides	Bisecting Diagonals	Equal Opposite Sides	Equal Opposite Angles	Equal Diagonals	All Sides Equal	All Angles Equal	Perpendicular Diagonals
Parallelogram	X	X	X	X				
Rectangle	X	X	X	X	X		X	
Rhombus	X	X	X	X		X		X
Square	X	X	X	X	X	X	X	X

Solving Problems Involving Special Quadrilaterals

The attributes of particular special quadrilaterals can be the key to solving problems that involve these figures. For instance, a problem might involve determining whether a quadrilateral with some given characteristics is a parallelogram. For such a problem, it is helpful to realize that the quadrilateral is a parallelogram if any *one* of the following statements is true:

1. One pair of opposite sides is both parallel and congruent

2. Both pairs of opposite sides are congruent

3. Both pairs of opposite angles are congruent

4. The diagonals bisect each other

Thus, it is not always necessary to demonstrate the applicability of every attribute of a special quadrilateral when attempting to identify a given instance. The following examples further illustrate the use of the characteristics of special quadrilaterals for solving problems.

Example: Find the measures of the other three angles of a parallelogram if one angle measures 38º.

Since opposite angles are equal, there are two angles measuring 38º. Also, since adjacent angles are supplementary, 180º − 38º = 142º, so the other two angles measure 142º each. To check:

$$
\begin{array}{r}
38 \\
38 \\
142 \\
+\ 142 \\
\hline
360
\end{array}
$$

Example: The measures of two adjacent angles of a parallelogram are 3x + 40 and x + 70. Find the measures of each angle.

$$
\begin{aligned}
2(3x + 40) + 2(x + 70) &= 360 \\
6x + 80 + 2x + 140 &= 360 \\
8x + 220 &= 360 \\
8x &= 140 \\
x &= 17.5 \\
3x + 40 &= 92.5 \\
x + 70 &= 87.5
\end{aligned}
$$

Thus, the pairs of opposite angles measure 92.5º and 87.5º each.

Example: Determine whether the following statements are true or false?

All squares are rhombuses.	True.
All parallelograms are rectangles.	False: a parallelogram need not have right interior angles.
All rectangles are parallelograms.	True.
Some rhombuses are squares.	True.
Some rectangles are trapezoids.	False: trapezoids have only one pair of parallel sides.
All quadrilaterals are parallelograms.	False: some quadrilaterals have non-parallel sides.
Not all squares are rectangles.	False: all squares are rectangles.
Some parallelograms are rhombuses.	True.

Example: In rhombus ABCD, side AB = 3x − 7 and side CD = x + 15. Find the length of each side.

Since all the sides are the same length, $3x − 7 = x + 15$

$$2x = 22$$
$$x = 11$$

Since $3(11) − 7 = 26$ and $11 + 15 = 26$, each side measures 26 units.

Example: In trapezoid ABCD, AB = 17 and DC = 21. Find the length of the median.

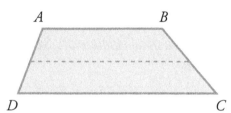

The median is one-half the sum of the bases.
$\frac{1}{2}(17 + 21) = 19$

Example: Given parallelogram ABCD with diagonals AC and BD intersecting at E, prove that AE ≅ CE.

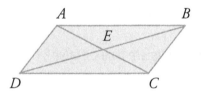

1.	Parallelogram *ABCD*, with diagonals *AC* and *BD* intersecting at *E*	Given.
2.	$AB \parallel DC$	Opposite sides of a parallelogram are parallel.
3.	$\angle BDC \cong \angle ABD$	If parallel lines are cut by a transversal, their alternate interior angles are congruent.
4.	$AB \cong DC$	Opposite sides of a parallelogram are congruent.
5.	$\angle BAC \cong \angle ACD$	If parallel lines are cut by a transversal, their alternate interior angles are congruent.
6.	$\triangle ABE \cong \triangle CDE$	ASA.
7.	$AE \cong CE$	Corresponding parts of congruent triangles are congruent.

Example: Given quadrilateral ABCD with AB ≅ DC and ∠BAC ≅ ∠ACD, prove that ABCD is a parallelogram.

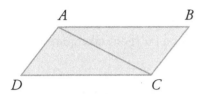

1.	Quadrilateral *ABCD* *AB* ≅ *DC* ∠*BAC* ≅ ∠*ACD*	Given.
2.	*AC* ≅ *AC*	Reflexive.
3.	△*ABC* ≅ △*ADC*	SAS.
4.	*AD* ≅ *BC*	Corresponding parts of congruent triangles are congruent.
5.	*ABCD* is a parallelogram	If both pairs of opposite sides of a quadrilateral are congruent, the quadrilateral is a parallelogram.

SKILL 2.9 Solve problems involving angles, diagonals, and vertices of polygons with more than four sides

Properties of Polygons

A POLYGON is a simple closed figure composed of line segments. Triangles and quadrilaterals, for instance, are polygons. The number of sides in a polygon is equal to the number of vertices; thus, for instance, a quadrilateral has four sides and four vertices. In a REGULAR POLYGON, all of the sides are the same length and all of the angles are the same measure.

The sum of the measures of the interior angles of a polygon can be determined using the following formula, where n represents the number of sides in the polygon:

Sum of angles = $180(n - 2)$

The measure of each angle of a regular polygon can be found by dividing the sum of the measures by the number of angles:

Measure of ∠ = $\frac{180(n - 2)}{n}$

> **POLYGON:** a simple closed figure composed of line segments

> **REGULAR POLYGON:** all of the sides are the same length and all of the angles are the same measure

Example: Find the measure of each angle of a regular octagon.

Since an octagon has eight sides, each angle equals

$$\frac{180(8-2)}{8} = \frac{180(6)}{8} = 135°$$

The sum of the measures of the exterior angles of a polygon, taken one angle at each vertex, equals 360°.

The measure of each exterior angle of a regular polygon can be determined using the following formula, where n represents the number of angles in the polygon:

Measure of exterior $\angle$ of regular polygon $= 180 - \frac{180(n-2)}{n}$

or, more simply, $= \frac{360}{n}$

Example: Find the measure of the interior and exterior angles of a regular pentagon.

Since a pentagon has five sides, each exterior angle measures

$$\frac{360}{5} = 72°$$

Since each exterior angle is supplementary to its interior angle, the interior angle measures $180 - 72$ or 108°.

Problems may also involve the vertices or diagonals of a polygon. For instance, a specific problem might call for the calculation of the area of a triangle formed by a diagonal in a polygon. In such a case, it is helpful to apply the characteristics of the polygon (especially if the problem involves a regular polygon). Consider the following example.

Example: Calculate the area of triangle ABC in the regular pentagon with sides of length 2.

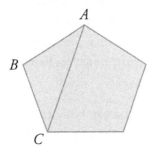

Since the sides of the polygon are all of length 2, triangle *ABC* is isosceles; angles $\angle BAC$ and $\angle BCA$ are therefore congruent. In addition, the interior angles of a regular pentagon are 108°. First, bisect this angle with an altitude, thereby forming two right triangles.

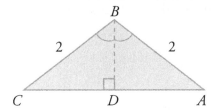

Since the bisected angle yields two 54° angles, angles $\angle BAC$ and $\angle BCA$ are each equal to 36°. Using the trigonometric ratios, the lengths CD and BD can be found. The area of triangle ABC is simply the product of CD and BD (or half the base times the height).

$$\cos 36° = \frac{CD}{2}$$
$$CD = 2\cos 36° \approx 1.618$$
$$\sin 36° = \frac{BD}{2}$$
$$BD = 2\sin 36° \approx 1.176$$

Thus, to an accuracy of three decimal places, the area of triangle ABC is (1.618)(1.176) = 1.903 square units.

> **SKILL 2.10** Solve problems that involve using the properties of circles, including problems involving inscribed angles, central angles, radii, tangents, arcs, and sectors

Properties of Circles

Central angles and arcs

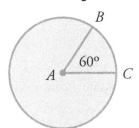

Central angle $BAC = 60°$
Minor arc $BC = 60°$
Major arc $BC = 360 - 60 = 300°$

If you draw two radii in a circle, the angle they form with the center as the vertex is a CENTRAL ANGLE. The piece of the circle "inside" the angle is an arc. Just like a central angle, an arc can have any degree measure from 0 to 360. The measure of an arc is equal to the measure of the central angle that forms the arc. Since a diameter forms a semicircle and the measure of a straight angle like a diameter is 180°, the measure of a semicircle is also 180°.

> **CENTRAL ANGLE:** when drawing two radii in a circle, this is the angle they form with the center as the vertex

Given two points on a circle, there are two different arcs that the two points form. Except in the case of semicircles, one of the two arcs will always be greater than

> *The measure of an arc is equal to the measure of the central angle that forms the arc.*

180°, and the other will be less than 180°. The arc less than 180° is a minor arc, and the arc greater than 180° is a major arc.

Examples:

1.

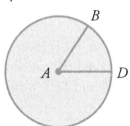

$m\angle BAD = 45°$

What is the measure of the major arc *BD*?

$\angle BAD$ = minor arc *BD* The measure of the central angle is the same as the measure of the arc it forms.

$45° = $ minor arc *BD*

$360 - 45 = $ major arc *BD* A major and minor arc always add to 360°.

$315° = $ major arc *BD*

2.

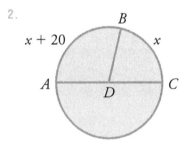

$\overline{AC}$ is a diameter of circle *D*. What is the measure of $\angle BDC$?

$m\angle ADB + m\angle BDC = 180°$ A diameter forms a semicircle that has
$2x + 20 = 180$ a measure of 180°.
$2x = 160$
$x = 80$

minor arc $BC = 80°$ A central angle has the same measure
$m\angle BDC = 80°$ as the arc it forms.

While an arc has a measure associated to the degree measure of a central angle, it also has a length, which is a fraction of the circumference of the circle.

For each central angle and its associated arc, there is a sector of the circle that resembles a pie piece. The area of such a sector is a fraction of the area of the circle.

The fractions used for the area of a sector and length of its associated arc are both equal to the ratio of the central angle to 360°.

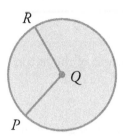

$$\frac{\angle PQR}{360°} = \frac{\text{length of arc } RP}{\text{circumference of } OQ} = \frac{\text{area of sector } PQR}{\text{area of } OQ}$$

Example:

1.

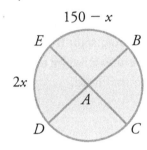

OA has a radius of 4 cm. What is the length of arc *ED*?

$2x + 150 - x = 180$

$x + 150 = 180$ Arc *BE* and arc *DE* make a semicircle.

$x = 30$

Arc $ED = 2(30) = 60$

$\frac{60}{360} = \frac{\text{arc length } ED}{2\pi 4}$ The ratio 60° to 360° is equal to the ratio of arc length *ED* to the circumference of *OA*. Cross-multiply and solve for the arc length.

$\frac{1}{6} = \frac{\text{arc length}}{8\pi}$

$\frac{8\pi}{6} = \text{arc length}$

Arc length $ED = \frac{4\pi}{3}$ cm.

2.

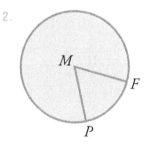

The radius *MP* is 3 cm. The length of arc *PF* is 2π cm. What is the area of sector *PMF*?

Circumference of circle $= 2\pi(3) = 6\pi$ Find the circumference and area of the circle.

Area of $\bigcirc M = \pi(3)^2 = 9\pi$ The ratio of the sector area to the circle area is the same as the arc length to the circumference.

$\frac{\text{area of } PMF}{9\pi} = \frac{2\pi}{6\pi}$

$$\frac{\text{area of } PMF}{9\pi} = \frac{1}{3}$$
$$\text{area of } PMF = \frac{9\pi}{3}$$
$$\text{area of } PMF = 3\pi \qquad \text{Solve for the area of the sector.}$$

Tangent lines and chords

A TANGENT LINE intersects a circle in exactly one point. If a radius is drawn to that point, the radius will be perpendicular to the tangent.

A CHORD is a segment with endpoints on the circle. If a radius or diameter is perpendicular to a chord, the radius will cut the chord into two equal parts.

If two chords in the same circle have the same length, the two chords will have arcs that are the same length, and the two chords will be equidistant from the center of the circle. Distance from the center to a chord is measured by finding the length of a segment from the center perpendicular to the chord.

Examples:

1.

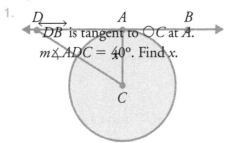

$\overrightarrow{DB}$ is tangent to $\odot C$ at A.
$m\angle ADC = 40°$. Find x.

$\overline{AC} \perp \overline{DB}$	A radius is $\perp$ to a tangent at the point of tangency.
$m\angle DAC = 90°$	Two segments that are $\perp$ form a 90° angle.
$40 + 90 + x = 180$	The sum of the angles of a triangle is 180°.
$x = 50$	Solve for x.

2.

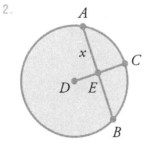

$\overline{CD}$ is a radius and $\overline{CD} \perp$ chord $\overline{AB}$.
$\overline{AB} = 10$. Find x.

$x = \frac{1}{2}(10)$ If a radius is $\perp$ to a chord, the radius bisects the chord.
$x = 5$

Angles

Angles with their vertices on a circle

An **INSCRIBED ANGLE** is an angle whose vertex is on the circle. Two chords, a diameter and a chord, two secants, or a secant and a tangent could form such an angle. An inscribed angle has one arc of the circle in its interior. The measure of the inscribed angle is one-half the measure of this intercepted arc. If two inscribed angles intercept the same arc, the two angles are congruent (i.e., their measures are equal). If an inscribed angle intercepts an entire semicircle, the angle is a right angle.

> **INSCRIBED ANGLE:** an angle whose vertex is on the circle

Angles with their vertices interior to a circle

When two chords intersect inside a circle, two sets of vertical angles are formed. Each set of vertical angles intercepts two arcs, which are across from each other. The measure of an angle formed by two chords in a circle is equal to one-half the sum of the angle intercepted by the angle and the arc intercepted by its vertical angle.

Angles with their vertices exterior to a circle

If an angle has its vertex outside of the circle and each side of the angle intersects the circle, then the angle contains two different arcs. The measure of the angle is equal to one-half the difference of the two arcs.

Example:

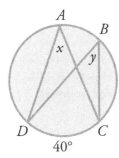

Find x and y.
arc $DC = 40°$

$m \angle DAC = \frac{1}{2}(40) = 20°$ $\angle DAC$ and $\angle DBC$ are both inscribed angles, so each one has a measure equal to one-half the measure of arc DC.

$m \angle DBC = \frac{1}{2}(40) = 20°$
$x = 20°$ and $y = 20°$

Intersecting chords

If two chords intersect inside a circle, each chord is divided into two smaller segments. The product of the lengths of the two segments formed from one chord equals the product of the lengths of the two segments formed from the other chord.

Intersecting tangent segments

If two tangent segments intersect outside of a circle, the two segments have the same length.

Intersecting secant segments

If two secant segments intersect outside a circle, a portion of each segment will lie inside the circle, and a portion (called the exterior segment) will lie outside the circle. The product of the length of one secant segment and the length of its exterior segment equals the product of the length of the other secant segment and the length of its exterior segment.

Tangent segments intersecting secant segments

If a tangent segment and a secant segment intersect outside a circle, the square of the length of the tangent segment equals the product of the length of the secant segment and its exterior segment.

Examples:

1.

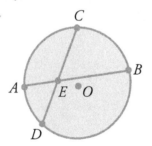

$\overline{AB}$ and $\overline{CD}$ are tangents.
$CE = 10$, $ED = x$, $AE = 5$, $EB = 4$

$(AE)(EB) = (CE)(ED)$ Since the chords intersect in the circle, the
$5(4) = 10x$ products of the segment pieces are equal.
$20 = 10x$
$x = 2$ Solve for x.

2.

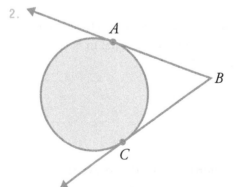

$\overline{AB}$ and $\overline{CD}$ are chords.
$\overline{AB} = x^2 + x - 2$
$\overline{BC} = x^2 - 3x + 5$
Find the lengths of $\overline{AB}$ and $\overline{BC}$

$\overline{AB} = x^2 + x - 2$ Given.

$$\overline{BC} = x^2 - 3x + 5$$
$$\overline{AB} = \overline{BC} \qquad \text{Intersecting tangents are equal.}$$
$$x^2 + x - 2 = x^2 - 3x + 5 \qquad \text{Set the expressions equal and solve.}$$
$$4x = 7$$
$$x = 1.75$$
$$(1.75)^2 + 1.75 - 2 = \overline{AB} \qquad \text{Substitute and solve.}$$
$$\overline{AB} = \overline{BC} = 2.81$$

Relationships Between Circles and Triangles

Inscribed and circumscribed circles

A circle is inscribed in a triangle if the three sides of the triangle are each tangent to the circle. The center of an inscribed circle is called the incenter of the triangle. To find the incenter, draw the three angle bisectors of the triangle. The point of concurrency of the angle bisectors is the incenter or center of the inscribed circle. Each triangle has only one inscribed circle.

A circle is circumscribed about a triangle if the three vertices of the triangle are all located on the circle. The center of a circumscribed circle is called the circumcenter of the triangle. To find the circumcenter, draw the three perpendicular bisectors of the sides of the triangle. The point of concurrency of the perpendicular bisectors is the circumcenter or the center of the circumscribing circle. Each triangle has only one circumscribing circle.

If two circles have radii that are in a ratio of $a : b$, then the following ratios are also true for the circles.

- The diameters are in the ratio of $a : b$

- The circumferences are in the ratio $a : b$

- The areas are in the ratio $a^2 : b^2$, or the ratio of the areas is the square of the ratio of the radii

Constructing a tangent to a circle

Given a circle with center O and a point on the circle such as P, construct the line tangent to the circle at P by constructing the line perpendicular to the radius drawn to P. If a line is perpendicular to a radius, the line will be tangent to the circle. For constructing a tangent to a circle, follow these steps.

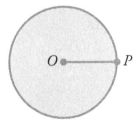

1. Draw the radius from *O* to P.

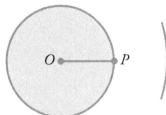

2. Open the compass from point *P* to point *O*. Use this radius to swing an arc from *P* to the exterior of the *P* circle.

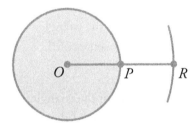

3. Put the straightedge on the radius and extend the segment to the arc forming segment *OR*. Note that *P* is the midpoint of *OR*.

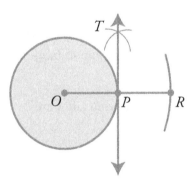

4. Open the compass to any length longer than the radius of this circle. Swing an arc of this radius from each endpoint of segment *OR*. Label the point where these two arcs intersect *T*.

5. Connect the point *P* and *T* to form the tangent line to circle *O* at point *P*.

Constructing an inscribed circle

For every triangle, there is one circle that is inscribed inside the triangle—in other words, the three sides of the triangle are tangent to the circle. The center of the inscribed circle is the point where the angle bisectors of the triangle meet. The three angle bisectors meet in a single point of concurrency called the incenter. Since the three angle bisectors meet in a single point, you only need to construct two of the angle bisectors to find the incenter. Follow these steps to find the inscribed circle for triangle *ABC*.

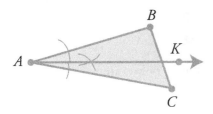

1. Construct the angle bisector of angle *A*. Label a point on the angle bisector point *K*, forming ray *AK*.

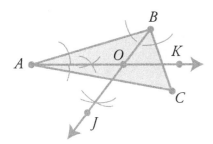

2. Construct the angle bisector of another angle—for example, angle *B*. Label a point on this bisector point *J*, forming ray *BJ*.

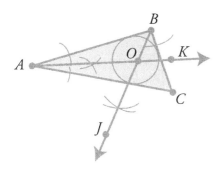

3. Label the point where the two angle bisectors meet point *O*. *O* is in the center of triangle *ABC*.

4. Place one end of the compass at point *O*, and open the compass so it stretches to one of the triangle sides. Verify that if the compass is held taut, it will just reach to the other two sides. If it will not reach to the other sides, readjust the compass until you find the one length that will reach all three sides. With this radius, swing a circle around point *O*.

SKILL 2.11 **Solve problems involving reflections, rotations, and translations of points, lines, or polygons in the plane**

Points, Lines, and Planes

In geometry, the point, line, and plane are key concepts and can be discussed in relation to one another.

Collinear points are all on the same line.

Noncollinear points are not on the same line.

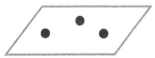

Coplanar points are on the same plane.

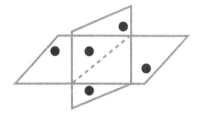

Noncoplanar points are not on the same plane.

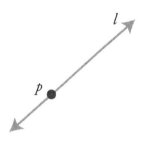

Point *p* is in line *l*.
Point *p* is on line *l*.
l contains *P*.
l passes through *P*.

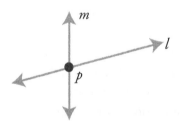

l and *m* intersect at *p*.
p is the intersection of *l* and *m*.

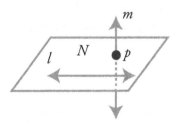

l and *p* are in plane *N*.
N contains *p* and *l*.
m intersects *N* at *p*.
p is the intersection of *m* and *N*.

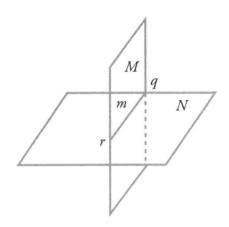

Planes *M* and *N* intersect at *rq*.
rq is the intersection of *M* and *N*.
rq is in *M* and *N*.
M and *N* contain *rq*.

Reflections

An object and its REFLECTION have the same shape and size, but the figures face in opposite directions. The line (where a hypothetical mirror may be placed) is called the LINE OF REFLECTION. The distance from a point to the line of reflection is the same as the distance from the point's image to the line of reflection.

REFLECTION: the figures have the same shape and size, but face in opposite directions

LINE OF REFLECTION: the line where a hypothetical mirror may be placed

Rotations

A ROTATION is a transformation that turns a figure about a fixed point, called the center of rotation. An object and its rotation are the same shape and size, but the figures may be turned in different directions. Rotations can occur in either a clockwise or a counterclockwise direction.

ROTATION: a transformation that turns a figure about a fixed point

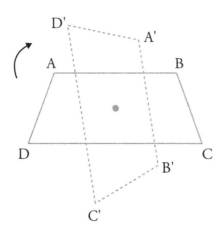

Translations

> **TRANSLATION:** a transformation that "slides" an object a fixed distance in a given direction

A TRANSLATION is a transformation that "slides" an object a fixed distance in a given direction. The original object and its translation have the same shape and size, and they face in the same direction.

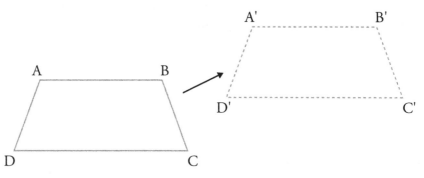

Translation is an intuitive concept. Movement of any object in real life from one position to another is an example of translation. So, for example, movement of a floor speaker from one place to another is a case of translation (assuming that the speaker is always pointed in the same absolute direction and that the floor is flat, thus avoiding any kind of rotation).

Problems Involving Transformations

Example: Plot the given ordered pairs on a coordinate plane and join them in the given order, then join the first and last points.

(-3, -2), (3, -2), (5, -4), (5, -6), (2, -4), (-2, -4), (-5, -6), (-5, -4)

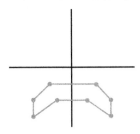

Increase all *y*-coordinates by 6.

(-3, 4), (3, 4), (5, 2), (5, 0), (2, 2), (-2, 2), (-5, 0), (-5, 2)

Plot the points and join them to form a second figure.

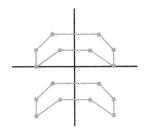

The examples of a translation, rotation, and reflection just given are for polygons, but the same principles apply to the simpler geometrical elements of points and lines. In fact, a transformation performed on a polygon can be viewed equivalently as the same transformation performed on the set of points (vertices) and lines (sides) that compose the polygon. Thus, to perform complicated transformations on a figure, it is helpful to perform the transformations on all the points (or vertices) of the figure and then reconnect the points with lines as appropriate.

Multiple transformations can be performed on a geometrical figure. The order of these transformations may or may not be important. For instance, multiple translations can be performed in any order, as can multiple rotations (around a single fixed point) or reflections (across a single fixed line). The order of the transformations becomes important when several types of transformations are performed or when the point of rotation or the line of reflection changes among transformations. For example, consider a translation of a given distance upward and a clockwise rotation by 90° around a fixed point. Changing the order of these transformations changes the result.

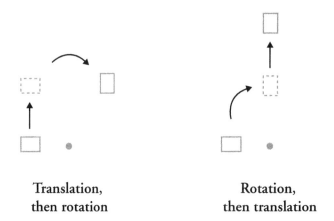

**Translation,
then rotation**

**Rotation,
then translation**

As shown, the final position of the box is different, depending on the order of the transformations. Thus, it is crucial that the proper order of transformations (whether determined by the details of the problem or some other consideration) be followed.

Example: Find the final location of a point at (1, 1) that undergoes the following transformations: rotate 90° counterclockwise about the origin; translate distance 2 in the negative y direction; reflect about the y-axis.
First, draw a graph of the *x*- and *y*-axes and plot the point at (1, 1).

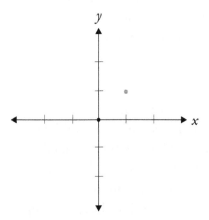

Next, perform the rotation. The center of rotation is the origin and is in the counterclockwise direction. In this case, the even value of 90° makes the rotation simple to do by inspection. Next, perform a translation of distance 2 in the negative y direction (down). The results of these transformations are shown below.

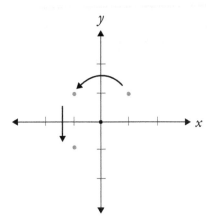

Finally, perform the reflection about the *y*-axis. The result, shown below, is a point at (1, –1).

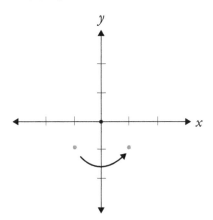

Using this approach, polygons can be transformed on a point-by-point basis.

For some problems, there is no need to work with coordinate axes. For instance, the problem may simply require transformations without respect to any absolute positioning.

Example: Rotate the following regular pentagon by 36° about its center and then reflect it about a horizontal line.

First, perform the rotation. In this case, the direction is not important because the pentagon is symmetric. As it turns out in this case, a rotation of 36° yields the same result as flipping the pentagon vertically (assuming the vertices of the pentagon are indistinguishable).

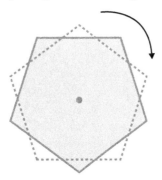

Finally, perform the reflection. Note that the result here is the same as a downward translation (assuming the vertices of the pentagon are indistinguishable).

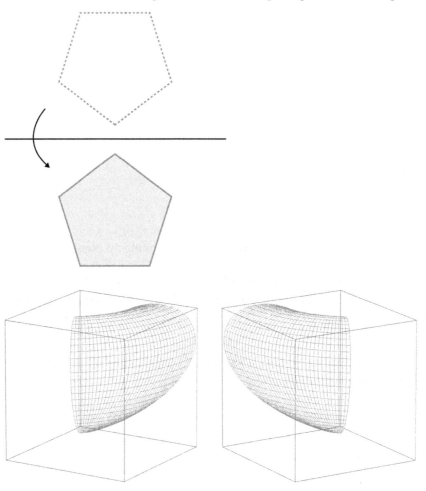

Above is an example of the rotation of a three-dimensional object. A 3D coordinate system consists of x-, y-, and z-axes that cross one another at the origin, and just like in a 2D graph, the position of an object is measured along each axis. 3D objects can be rotated, reflected, translated, and dilated just like two-dimensional objects. To translate a 3D point, you can change each dimension separately:

$x' = x + a1$

$y' = y + a1$

$z' = z + a1$

Below is a representation of a cube on the *xyz*-axes.

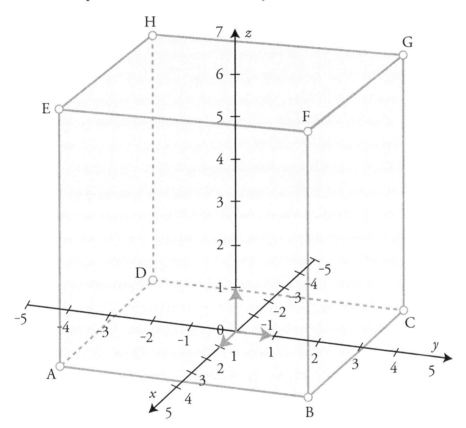

For more information on 3D manipulations, visit:

http://demonstrations.
wolfram.com/Uderstand-
ing3DTranslation

SKILL Solve problems that can be represented on the XY-plane *(e.g., finding*
2.12 *the distance between two points or finding the coordinates of the midpoint of a line segment)*

The Distance Formula

In order to accomplish the task of finding the distance from a given point to another given line, the perpendicular line that intersects the point and line must be drawn, and the equation of the other line must be written. From this information, the point of intersection can be found. This point and the original point are used in the distance formula given below:

$$D = \sqrt{(x_2 - x_1)^2 + (y_2 - y_1)^2}$$

Example: Given the point (-4, 3) and the line y = 4x + 2, find the distance from the point to the line.

$y = 4x + 2$	Find the slope of the given line by solving for y.
$y = 4x + 2$	The slope is $\frac{4}{1}$; the perpendicular line will have a slope of $-\frac{1}{4}$.
$y = (-\frac{1}{4})x + b$	Use the new slope and the given point to find the equation of the perpendicular line.
$3 = (-\frac{1}{4})(-4) + b$	Substitute (-4, 3) into the equation.
$3 = 1 + b$	Solve.
$2 = b$	Given the value for b, write the equation of the perpendicular line.
$y = (-\frac{1}{4})x + 2$	Write in standard form.
$x + 4y = 8$	Use both equations to solve by elimination to get the point of intersection.
$-4x + y = 2$	Multiply the bottom row by 4.
$\underline{x + 4y = 8}$	

$$-4x + y = 2 \qquad \text{Solve.}$$
$$\underline{4x + 16y = 32}$$
$$17y = 34$$
$$y = 2$$

$y = 4x + 2$	Substitute to find the x value.
$2 = 4x + 2$	Solve.
$x = 0$	

(0, 2) is the point of intersection. Use this point on the original line and the original point to calculate the distance between them.

$D = \sqrt{(x_2 - x_1)^2 + (y_2 - y_1)^2}$, where points are (0, 2) and (-4, 3).

$D = \sqrt{(-4 - 0)^2 + (3 - 2)^2}$ Substitute.

$D = \sqrt{(16) + (1)}$ Simplify.

$D = \sqrt{17}$

The distance between two parallel lines, such as line *AB* and line *CD* as shown below, is the line segment *RS*, the perpendicular between the two parallels.

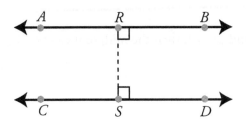

Example: Given the geometric figure below, find the distance between the two parallel sides AB and CD.

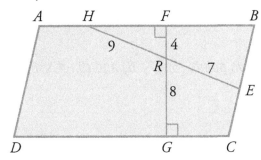

The distance *FG* is 12 units.

The key to applying the distance formula is to understand the problem before beginning.

$$D = \sqrt{(x_2 - x_1)^2 + (y_2 - y_1)^2},$$

Example: Find the perimeter of a figure with vertices at (4, 5), (-4, 6), and (-5, -8).

The figure being described is a triangle. Therefore, the distance for all three sides must be found. Carefully identify all three sides before beginning.

    Side 1 = (4, 5) to (-4, 6)
    Side 2 = (-4, 6) to (-5, -8)
    Side 3 = (-5, -8) to (4, 5)

$$D_1 = \sqrt{(-4 - 4)^2 + (6 - 5)^2} = \sqrt{65}$$
$$D_2 = \sqrt{((-5 - (-4))^2 + (-8 - 6)^2} = \sqrt{197}$$
$$D_3 = \sqrt{((4 - (-5))^2 + (5 - (-8))^2} = \sqrt{250} \text{ or } 5\sqrt{10}$$

$$\text{Perimeter} = \sqrt{65} + \sqrt{197} + 5\sqrt{10}$$

The Midpoint Formula

If a line segment has endpoints of (x_1, y_1) and (x_2, y_2), then the midpoint can be found using

$$\left(\frac{x_1 + x_2}{2}, \frac{y_1 + y_2}{2}\right)$$

Example: Find the center of a circle with a diameter whose endpoints are (3, 7) and (-4, -5).

$$\text{Midpoint} = \left(\frac{3 + (-4)}{2}, \frac{7 + (-5)}{2}\right)$$

$$\text{Midpoint} = \left(\frac{-1}{2}, 1\right)$$

Example: Find the midpoint given the two points (5, 8$\sqrt{6}$) and (9, -4$\sqrt{6}$).

$$\text{Midpoint} = \left(\frac{5 + 9}{2}, \frac{8\sqrt{6} + (-4\sqrt{6})}{2}\right)$$

$$\text{Midpoint} = (7, 2\sqrt{6})$$

> **SKILL 2.13** Estimate absolute and relative error in the numerical answer to a problem by analyzing the effects of round-off and truncation errors introduced in the course of solving a problem

Round-off and Truncation Error

Problems that involve calculating numerical quantities are often subject to error due to a lack of infinite precision in the calculator or on the part of the person doing the calculations. For instance, calculations that involve the number π cannot be written in decimal form without either round-off or truncation of the answer. For instance:

$$1.5\pi = 4.71238898\ldots$$

The earlier in a series of calculations that inexact decimal values are introduced, the more serious the error can become at later stages of the calculation. The transmission of numerical error through a series of calculations is called PROPAGATION OF ERROR. Consider the following series of calculations and the effect that using decimal values has when the results are rounded off to three decimal places. In each column, rounded decimal values are used earlier in the series, and the effect of round-off is shown as it progresses through the calculations.

PROPAGATION OF ERROR: the transmission of numerical error through a series of calculations

Step 1	$\theta = \frac{\pi}{4}$	$\theta = \frac{\pi}{4}$	$\theta = \frac{\pi}{4}$	$\theta = \frac{\pi}{4}$
Step 2	$\alpha = \sin\theta$	$\alpha = \sin\theta$	$\alpha = \sin\theta$	$\alpha = 0.014$
Step 3	$\beta = \ln(\alpha)$	$\beta = \ln(\alpha)$	$\beta = -4.290$	$\beta = -4.269$

| **Step 4** | β^2 | 18.403 | 18.404 | 18.224 |

As demonstrated above, the sooner a decimal value of limited precision is used, and the sooner that round-off error begins propagating through the calculations, the more drastic the effect on the resulting answer.

The exact effect of round-off or truncation on the error of the final result of a series of calculations is not always easy to predict. In some cases, depending on the particular numbers used and how significant the portion of the number that is removed by round-off or truncation, the effect may be insignificant. In the example above, beginning to use a rounded decimal value in Step 3 (instead of Step 4) has little effect, but beginning to use rounded decimal values in Step 2 causes significant deviation.

One way to estimate the effect of round-off or truncation on the results of a calculation or series of calculations is to determine how close the rounded number is to the original number. For instance, rounding 1.5 to 2 is a significant change—about a 33% change in the original value. A simple multiplication by a factor, such as 4, would yield two drastically different answers: 6 for the unrounded number and 8 for the rounded number. This is also a difference of about 33%. (It is noteworthy that the percent change in results due to rounding at some stage in the calculation is not always identical to the percent change in the value that is rounded.) On the other hand, rounding 1.4999 to 1.5 is a change of less than 0.01%. Multiplication by 4 yields either 5.9996 for the unrounded number or 6 for the rounded number. If the answers are also rounded to one decimal place, the results are the same: 6.

As indicated above, the effect of round-off or truncation error also depends on the point in the problem at which the rounded or truncated values are inserted. Thus, a problem that is solved with a rounded number transferred from each calculation to another is likely to be less accurate than cases where only the final result is rounded. (Again, the rounding or truncation can take place deliberately by the person doing the calculations, or it can be the inadvertent result of limited precision on the part of a calculator or computer.)

Short of deriving rules for the propagation of error as a function of the different types of calculations used (trigonometric functions, exponents, logarithms, addition and subtraction, etc.), the best way to test the potential range of numerical error for a problem is to use some different truncated or rounded values and examine their effect on the final solution. The greater the range of final results, the higher the potential for numerical error because of rounding or truncating.

Example: Estimate the numerical error due to rounding for the calculation of a^2, where a is 1.256.

First, note that a can be rounded to 1.26, 1.3, or 1. Calculate a^2 for each case:

$1.256^2 = 1.577536$

$1.26^2 = 1.5876$

$1.3^2 = 1.69$

$1^2 = 1$

Thus, the numerical error due to rounding could be as high as 0.577536, or about 37%.

SKILL **Demonstrate an intuitive understanding of a limit**
2.14

Graphical Interpretation of Limit

LIMIT: the value that a function approaches as a variable of the function approaches a certain value

Informally, a LIMIT is the value that a function approaches as a variable of the function approaches a certain value. This concept can be understood through graphical illustrations, as with the following example. The graph shows a plot of $y = 2x$. A limit in this case might be the limit of y as x approaches 3. Graphically, this involves tracing the plot of the function to the point at $x = 3$.

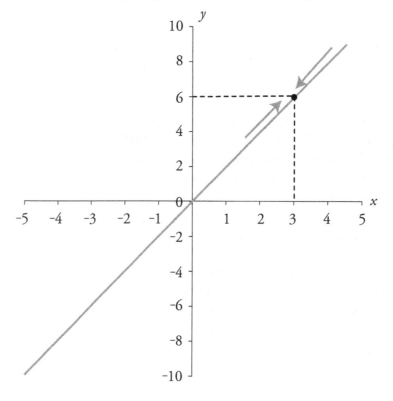

The limit in this case is clearly 6: $2(3) = 6$. The limit can be considered either

from the right or left (that is, considering x decreasing toward 3 or x increasing toward 3). In the simple example above, the limit is the same from either direction, but, in some cases, the limit may be different from different directions or the function may not even exist on one side of the limit. Consider the limit of ln x as x approaches zero.

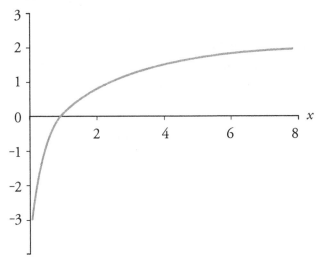

As it turns out, ln x is not defined for $x < 0$. The limit, however, is negative infinity (which could be called "undefined" as well). Nevertheless, it is clear that the graphical concept of a limit only makes sense for x approaching zero from the right, since the function does not exist (for real numbers) to the left of zero.

Numerical Interpretation of Limit

A limit may also be considered in a numerical sense. The limit is the value of the function as the variable (x, for instance) is increased or decreased incrementally towards the limiting value. For example, consider once again the limit of $y = 2x$ as x approaches 3.

INCREASING X		DECREASING X	
x	**y**	**x**	**y**
1.0	2.0	5.0	10.0
2.0	4.0	4.0	8.0
2.5	5.0	3.5	7.0
2.9	5.8	3.1	6.2
2.99	5.98	3.01	6.02

Table continued on next page

2.9999	5.9998	3.0001	6.0002
3.0	6.0	3.0	6.0

Again, the limit is the same regardless of the direction taken (increasing or decreasing x). Even without directly calculating the value of y for $x = 3$, it can be seen for other values that the value of the limit is apparently 6.

The concept of limit is a foundational principle of calculus, and having a solid understanding of limits is therefore critical to successfully advancing into higher-level mathematics.

<table>
<tr><td>SKILL 2.15</td><td>Demonstrate an intuitive understanding of maximum and minimum</td></tr>
</table>

Maxima and minima can exist for both discrete and continuous sets of numbers. In some cases, a maximum or minimum (or both) may not exist at all. For instance, the set of real numbers has no maximum or minimum value (unless positive and negative infinity are considered "values").

The concept of extrema (maxima and minima) can be differentiated into local and global versions as well. For instance, consider the following function:

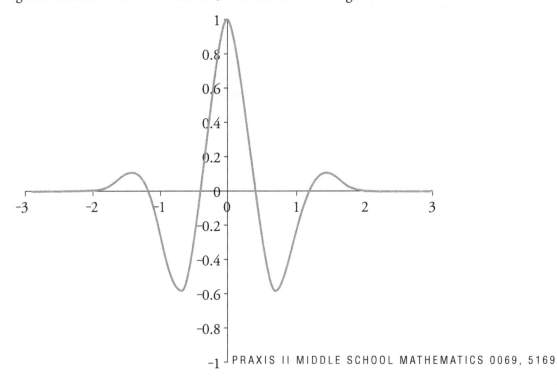

It is apparent that there are a number of peaks and valleys, each of which could, in some sense, be called a maximum or minimum. To allow greater clarity, local and global extrema can be specified. For instance, the peak at $x = 0$ is the maximum for the entire function. Additionally, the valleys at about $x = \pm 0.7$ both correspond to an (equivalent) minimum for the entire function. These are global extrema. On the other hand, the peaks at about $x = \pm 1.4$ are each a maximum for the function within a specific area; thus, they are local maxima.

The above case deals with extrema in the case of a continuous function. Maxima and minima can also be found for discrete functions or discrete sets. If the set is not ordered, there can be no concept of "local" extrema, but this concept can be applied if the set is ordered. For instance, the above graph could involve a discrete set of points instead of a continuous set.

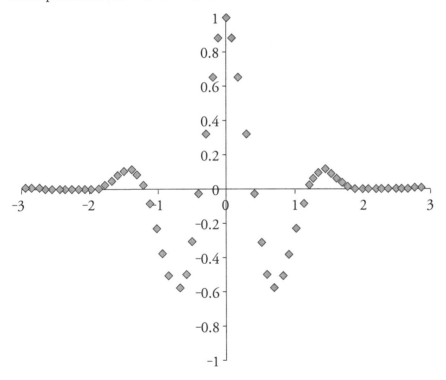

In this case, local and global extrema can still be identified. If the set of numbers is not ordered in the same sense as the above graph (consider, for instance, the set of prime numbers), then there can be no local extrema.

Prime numbers = {2, 3, 5, 7, 11, 13, 17, 19, 23,…}

On the other hand, the set of prime numbers does have a global minimum: 2. There is no defined global maximum (as far as is known), however.

A real-world example of extrema can be seen in geography. A particular mountain, for instance, may be the highest peak within sight (a local maximum), but it may not be as high as Mt. Everest (a global maximum). The same logic applies to the depth of valleys.

SKILL **Estimate the area of a region in the *xy*-plane**
2.16

There are a number of strategies for estimating the area of a region in the *xy*-plane. Using calculus, the exact area of the region can usually be found when simpler methods will not suffice, but simpler methods may often be sufficient for the purposes of estimation. In some cases, they may even be faster than calculus-based methods for arriving at an exact answer.

The approach used in calculus is one potential approach to estimation of the area of a region. This approach involves dividing the region into a set of rectangular areas that approximately covers the region. Finding the area of a rectangle involves a simple, well-known calculation, and calculating the estimate of the area of the region in the *xy*-plane is simply a matter of summing the areas of all the rectangles. For a very rough estimate, it is sufficient to use a small number of larger rectangles; for a more exact estimate, it is necessary to use a larger number of smaller rectangles.

Example: Estimate the area between the function and the x-axis.

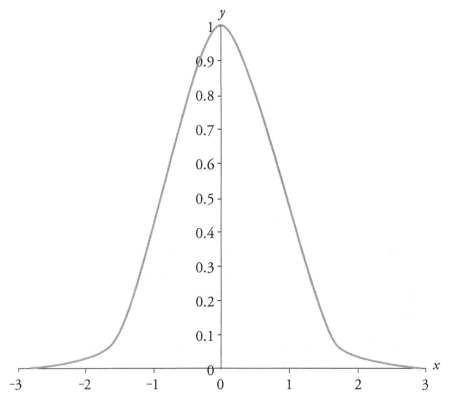

The area between the function and the x-axis can be estimated using several rectangles, as shown below. One of the keys to this approach is (by inspection)

making sure that the area of interest that is outside of the rectangles is approximately equal to the undesired area inside the rectangles, thus providing some balance in the error of the estimate.

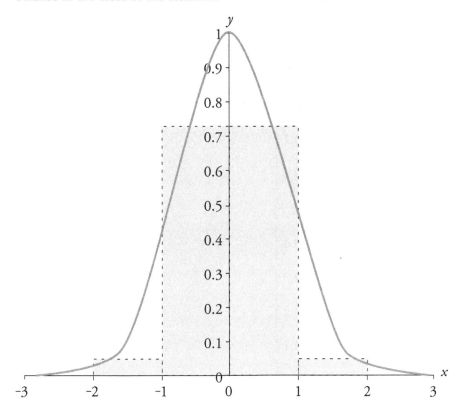

The area under the curve, based on this use of four rectangles, is approximately 1.55 square units. This estimate comes from adding two rectangles of height 0.05 units and width 1 unit (outer rectangles) and two rectangles of height 0.725 units and width 1 unit (inner rectangles). The actual area under the curve is approximately 1.77 square units.

By using a larger number of thinner rectangles, the same type of estimate can be made with even greater precision.

For this type of problem, other polygons of known area can be used in place of rectangles. Consider the same function as was used in the above example.

Example: Estimate the area between the function and the x-axis.

In this case, instead of using rectangles, use two triangles (since the two sides of the function are nearly straight over most of the domain of interest).

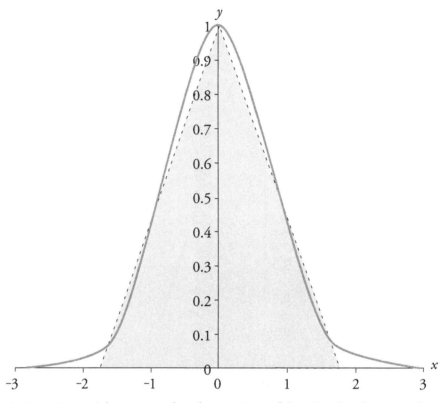

Again, it is crucial to ensure that the portions of the triangles that miss the area under the function are balanced by the area under the function that is outside the triangles. This helps reduce error in the estimate, and an inspection-based guess can be reasonably accurate.

The area of each triangle above is approximately $\frac{1}{2}(1)(1.75)$, or 0.875 square units. The sum of the area in both triangles is 1.75 square units, which is very close to the actual answer.

The same type of approach can be used for estimating the area in other types of plots in the *xy*-plane.

Example: Estimate the area inside the following figure.

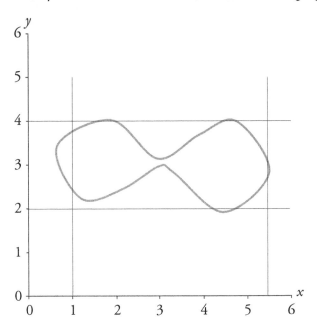

In this case, the figure shown has two nearly circular lobes. Estimate by using circles, as shown below.

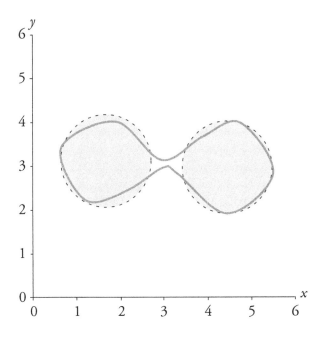

By inspection, it can be seen that the circles both have a diameter of about 1.8 units. The total area is then twice the product of π and the radius (half the diameter) squared. The area of a circle is 0.81π, or about 2.54 square units. The area of the figure can then be estimated as twice this value, or about 5.08 square units.

DOMAIN III
FUNCTIONS AND THEIR GRAPHS

PERSONALIZED STUDY PLAN

KNOWN MATERIAL/ SKIP IT

PAGE	COMPETENCY AND SKILL		KNOWN MATERIAL/ SKIP IT
153	**3:**	**Functions and their graphs**	☐
	3.1:	Understand function notation and the algebraic definition of a function	☐
	3.2:	Identify whether a graph in the plane is the graph of a function	☐
	3.3:	Given a graph select an equation that best represents the graph	☐
	3.4:	Determine the graphical properties and sketch a graph of a linear, step, absolute-value, quadratic, or exponential function	☐
	3.5:	Demonstrate an understanding of a physical situation or a verbal description of a situation and develop a model of it	☐
	3.6:	Determine whether a particular mathematical model, can be used to describe two seemingly different situations	☐
	3.7:	Find the domain (x-values) and range (y-values) of a function	☐
	3.8:	Translate verbal expressions and relationships into algebraic expressions or equations	☐

COMPETENCY 3
FUNCTIONS AND THEIR GRAPHS

Function Notation

A FUNCTION is a relation in which each value in the domain (i.e., x value) corresponds to only one value in the range (i.e., y value). It is notable, however, that a value in the range may correspond to any number of values in the domain. Functions of one variable are typically written in the following form, where y represents the function and x represents the variable:

$y(x)$

In some cases, where it is implicitly known that the function depends on a specific variable such as x, the function may simply be written as y instead of $y(x)$. The algebraic description of the function typically follows in the form of an equation, as with the following example.

$y(x) = x^2$ or $y = x^2$

This notation for a function indicates that the y value for a given x value is equal to (in this case) x^2.

It is important to notice that the particular symbols used in function notation are not important, as long as they are used consistently. Thus, the following are all the same functions, but simply use different symbols to represent the erstwhile x and y notation.

$f(\alpha) = \alpha^2$
$\beta(r) = r^2$
$\Pi(\square) = \square^2$

Determining if a Relation Is a Function

The fundamental principle for this function notation is to represent that for each x (or, generally, variable) value, the function (be it f, β, or any other symbol) has only one value. Thus, relations or equations that can have two different y values are not functions.

The following example is *not* a function, because *y* can have two different values for each *x* in the domain:

$$y(x) = \pm \sqrt{x}$$

Thus, to test whether a particular relation is a function, it is necessary to determine whether, for each *x* value, there is only one *y* value.

Example: Determine whether the following relations are functions:
$$y(x) = 2x^2 + x, \, y(\theta) = \sin\theta, \, y(\lambda) = \ln\lambda \pm 2.$$

The first two cases are both functions because, for each *x* or θ value, the function can have only one possible *y* value. (Determining this fact for some relations may require looking at a graph of the relation to determine whether the trends of the plot indicate the absence of multiple *y* values for certain *x* values.) In the third case, however, *y* can have two values for a given λ value:

$$y(1) = \ln 1 \pm 2 = 0 \pm 2 = \pm 2$$

Thus, $y(\lambda) = \ln\lambda \pm 2$ is not a function.

The following are some terms associated with relations and functions:

- A RELATION is any set of ordered pairs.

- The DOMAIN of a relation is the set of all the first coordinates of the ordered pairs.

- The RANGE of a relation is the set of all the second coordinates of the ordered pairs.

- A FUNCTION is a relation in which different ordered pairs have different first coordinates (no *x* values are repeated).

- A MAPPING is a diagram with arrows drawn from each element of the domain to the corresponding element(s) of the range. If two arrows are drawn from the same element of the domain, then the relation is not a function.

Example: Determine the domain and range of this mapping.

Answers

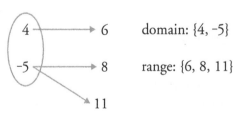

domain: {4, –5}

range: {6, 8, 11}

Example: If E = {(x, y) | y = 5}, find the domain and range.
Domain = (–∞, ∞)   Range = 5

RELATION: any set of ordered pairs

DOMAIN: the set of all the first coordinates of the ordered pairs

RANGE: the set of all the second coordinates of the ordered pairs

FUNCTION: a relation in which different ordered pairs have different first coordinates (no x values are repeated).

MAPPING: a diagram with arrows drawn from each element of the domain to the corresponding element(s) of the range

Example: Determine the ordered pairs of the relation shown in this mapping.

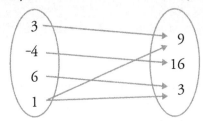

Answer: (3, 9), (–4, 16), (6, 3), (1, 9), (1, 3)

Example: Determine the domain and range of these graphs.

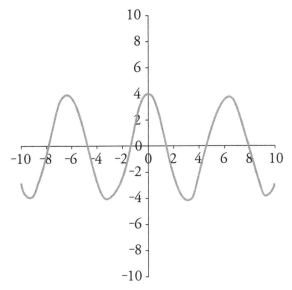

Domain = (–∞, ∞)   Range = –4, 4

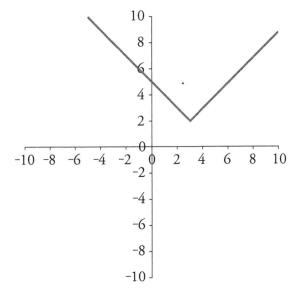

Domain = (–∞, ∞)   Range = 2, ∞

Identify whether a graph in the plane is the graph of a function; given a set of conditions, decide if they determine a function

The Vertical Line Test

The simplest way to determine whether a graph of a relation in a plane is the graph of a function is to use the VERTICAL LINE TEST. A vertical line on a graph (where the *y*-axis is also vertical and the *x*-axis is horizontal) connects all the possible *y* values for a given *x* value. Thus, if a vertical line crosses the plot of a relation just once for any given *x* value, then the graph is a function.

> **VERTICAL LINE TEST:** if a vertical line crosses the plot of a relation just once for any given x value, then the graph is a function

Example: Determine whether the following graph depicts a function.

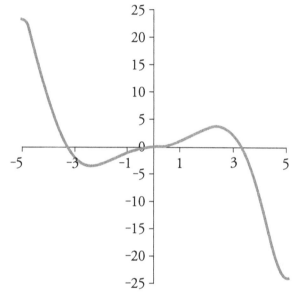

Use the vertical line test on the graph, as shown below. For every location of the vertical line, the plotted curve crosses the line only once, thus indicating that for every *x* value, there is only one *y* value. Thus, the graph depicts a function.

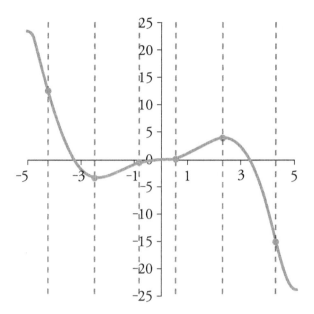

Example: Determine whether the following graph depicts a function.

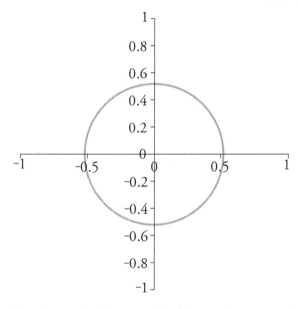

Use the vertical line test. In this case, however, the vertical line intersects the curve in more than one place, thus indicating there can be multiple *y* values for some *x* values. As a result, this graph does not meet the definition of a function.

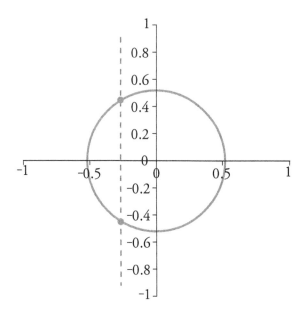

Determining a Function

By applying the definition of a function, it is possible to discover whether a set of conditions determines a function. The key is to keep in mind that for each *x* value (or variable value), a function produces a single *y* value (or function value). The set of conditions may involve if-then statements or other expressions that do not necessarily involve typical function notation.

Example: Decide if the following condition determines a function: For the set of natural numbers, if a number is prime, then it corresponds to 1; otherwise, it corresponds to 0.

In this case, the natural numbers (1, 2, 3, etc.) are the domain of a relation that converts them into either 1 for prime numbers or 0 for non-prime numbers. A graph of this relation might look like the following:

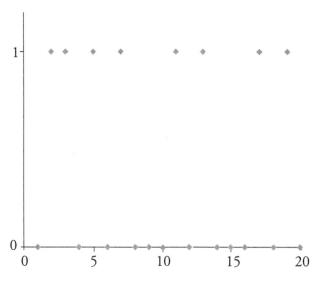

If the condition is plotted in this manner, the vertical line test can be applied. Otherwise, it is sufficient to note that for any natural number, the number is either prime or not prime, but not both. Thus, the condition can only yield exclusively 1 or 0 for a given number. As a result, for any x value (natural numbers), there is only one y value (1 or 0); this condition thus defines a function.

Example: Decide if the following condition determines a function: If a is a positive real number, then output the value of x for a = x².

In this case, the domain of the condition is the set of positive real numbers expressed as a. The relation yields the value of x for the equation $a = x^2$. Graphing the results is possible, but it is sufficient to simply solve for x. This yields $x = \pm \sqrt{a}$. Note that both positive and negative values of x can satisfy the equation $a = x^2$. As a result, each a value corresponds to two distinct x values (for instance, $(-2)^2 = (2)^2 = 4$ where $a = 4$).

This condition, then, does not determine a function.

Example: If A = {(x,y) | y = x² − 6}, find the domain and range.
 Answer: Domain = $(-\infty, \infty)$ Range = $-6, \infty$

Example: Give the domain and range of set B if:
 B = {(1,-2), (4,-2), (7,-2), (6,-2)}
 Answer: Domain = 1, 4, 7, 6   Range = -2

Example: Determine the domain of this function:
 $f(x) = \frac{5x + 7}{x^2 - 4}$
 Answer: Domain = $x \neq 2, -2$

Finding Values of a Function Using Substitution and Synthetic Division

There are several ways to find the values of a function. First, to find the value of a function $f(x)$ when $x = 3$, substitute 3 in place of every letter x. Then simplify the expression following the order of operations. For example, if $f(x) = x^3 - 6x + 4$, then to find $f(3)$, substitute 3 for x.

The equation becomes $f(3) = 3^3 - 6(3) + 4 = 27 - 18 + 4 = 13$.

So (3, 13) is a point on the graph of $f(x)$.

A second way to find the value of a function is to use synthetic division. To find the value of a function when $x = 3$, divide 3 into the coefficients of the function.

(Remember that coefficients of missing terms, like x^2, must be included). The remainder is the value of the function.

If $f(x) = x^3 - 6x + 4$, then to find $f(3)$ using synthetic division:

Note the 0 for the missing x^2 term.

$$3 \,\begin{array}{|rrrr} 1 & 0 & -6 & 4 \\ & 3 & 9 & 9 \\ \hline 1 & 3 & 3 & 13 \end{array} \leftarrow \text{This is the value of the function.}$$

Therefore, (3, 13) is a point on the graph of $f(x) = x^3 - 6x + 4$.

Example: Find values of the function at integer values from x = -3 to x = 3 if f(x) = x³ − 6x + 4.

If $x = -3$:

$f(-3) = (-3)^3 - 6(-3) + 4$

$\quad = (-27) - 6(-3) + 4$

$\quad = -27 + 18 + 4 = -5$

Synthetic division:

$$-3 \,\begin{array}{|rrrr} 1 & 0 & -6 & 4 \\ & -3 & 9 & -9 \\ \hline 1 & -3 & 3 & -5 \end{array} \leftarrow \text{This is the value of the function if } x = -3.$$

Therefore, (-3, -5) is a point on the graph.

If $x = -2$:

$f(-2) = (-2)^3 - 6(-2) + 4$

$\quad = (-8) - 6(-2) + 4$

$\quad = -8 + 12 + 4 = 8 \leftarrow$ This is the value of the function if $x = -2$.

Therefore, (-2, 8) is a point on the graph.

If $x = -1$:

$f(-1) = (-1)^3 - 6(-1) + 4$

$\quad = (-1) - 6(-1) + 4$

$\quad = -1 + 6 + 4 = 9$

Synthetic division:

$$-1 \,\begin{array}{|rrrr} 1 & 0 & -6 & 4 \\ & -1 & 1 & 5 \\ \hline 1 & -1 & -5 & 9 \end{array} \leftarrow \text{This is the value of the function if } x = -1.$$

Therefore, (-1, 9) is a point on the graph.

If $x = 0$:

$$f(0) = (0)^3 - 6(0) + 4$$
$$= 0 - 6(0) + 4$$
$$= 0 - 0 + 4 = 4 \leftarrow \text{This is the value of the function if } x = 0.$$

Therefore, $(0, 4)$ is a point on the graph.

If $x = 1$:

$$f(1) = (1)^3 - 6(1) + 4$$
$$= (1) - 6(1) + 4$$
$$= 1 - 6 + 4 = -1$$

Synthetic division:

$$
\begin{array}{r|rrrr}
 & 1 & 0 & -6 & 4 \\
1 & & 1 & 1 & -5 \\
\hline
 & 1 & 1 & -5 & -1
\end{array}
\leftarrow \text{This is the value of the function at } x = 1.
$$

Therefore, $(1, -1)$ is a point on the graph.

If $x = 2$:

$$f(2) = (2)^3 - 6(2) + 4$$
$$= 8 - 6(2) + 4$$
$$= 8 - 12 + 4 = 0$$

Synthetic division:

$$
\begin{array}{r|rrrr}
 & 1 & 0 & -6 & 4 \\
2 & & 2 & 4 & -4 \\
\hline
 & 1 & 2 & -2 & 0
\end{array}
\leftarrow \text{This is the value of the function at } x = 2.
$$

Therefore, $(2, 0)$ is a point on the graph.

If $x = 3$:

$$f(3) = (3)^3 - 6(3) + 4$$
$$= 27 - 6(3) + 4$$
$$= 27 - 18 + 4 = 13$$

Synthetic division:

$$
\begin{array}{r|rrrr}
 & 1 & 0 & -6 & 4 \\
3 & & 3 & 9 & 9 \\
\hline
 & 1 & 3 & 3 & 13
\end{array}
\leftarrow \text{This is the value of the function at } x = 3.
$$

Therefore, $(3, 13)$ is a point on the graph.

The following points are points on the graph:

X	-3	-2	-1	0	1	2	3
Y	-5	8	9	4	-1	0	13

Note the change in sign of the y value between $x = -3$ and $x = -2$. This indicates that there is a zero between $x = -3$ and $x = -2$. Since there is another change in sign of the y value between $x = 0$ and $x = -1$, there is a second root there. When $x = 2$, $y = 0$ so $x = 2$ is an exact root of this polynomial.

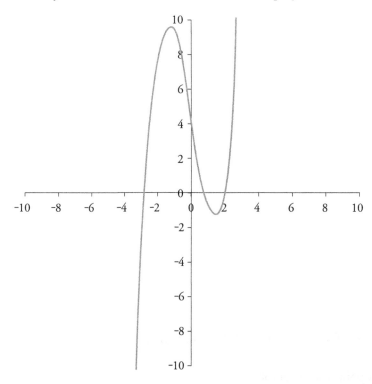

SKILL **Given a graph** (e.g, a line, a parabola, a step, absolute value, or simple
3.3 exponential) **select an equation that best represents the graph; given an equation, show an understanding of the relationship between the equation and its graph**

The first step in identifying the equation that best represents a given graph is first to determine the general type of curve that the graph represents. Typically, lines, parabolas, step functions, absolute value functions, and simple exponential functions are all easily distinguishable by virtue of their general shape. The following illustrative graphs show instances of several common types of functions.

Graphs of Common Functions

Lines

The graph of a line is easily recognizable. There are no discontinuities or other changes in the slope of the plot. The general equation for a line is $y = mx + b$. In the case of this example, the equation is $y = 2x$.

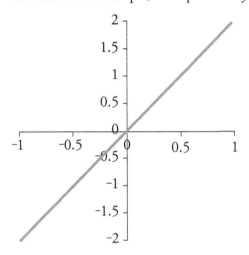

Step functions

A step function involves two horizontal lines ($y = a$, where a is some real number) with a sharp change (discontinuity) at some x value. In this case, the step function is defined as

$$y = \begin{cases} 1 & x \geq 0.3 \\ 0 & x < 0.3 \end{cases}$$

Notice that there is a closed point to indicate that at $x = 0.3$, $y = 1$ (rather than $y = 0$, which is further indicated by the presence of the open point).

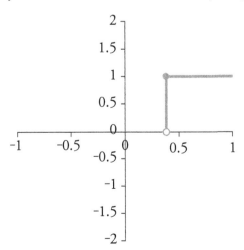

Absolute value functions

An absolute value function involves two rays, each with its own slope, that join at a point. Thus, this type of function looks like a sharply bent line. If multiple absolute value terms are included, there may be several different slope discontinuities. In this case, the equation of the absolute value function is $y = |x - 0.5| + 0.5$.

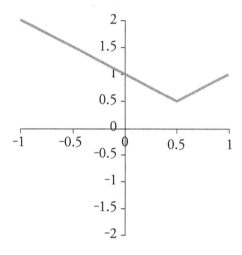

Quadratic functions

A parabola results from a quadratic function and has the curved shape shown in the example below. Parabolas have an obvious minimum or maximum. The concavity of the parabola can be directed up or down, depending on the sign of the quadratic term. Quadratic functions generally have the form $y = ax^2 + bx + c$, where a, b, and c are real numbers and $a \neq 0$. The parabola below is for the function $y = 2x^2 - x$.

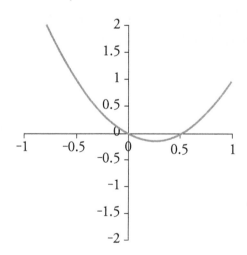

Exponential functions

Simple exponential functions plot as curves that increase steadily in steepness and that have no maximum or minimum. The general form of an exponential function is $y = ae^{bx}$, where a and b are real numbers. The example exponential function below represents $y = e^x$.

Other functions can be constructed that involve combinations of the above functions, trigonometric functions, polynomials of arbitrary degree and other more exotic functions.

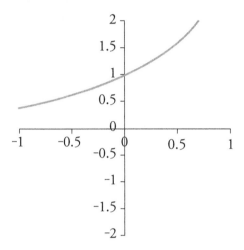

Finding the Equation of a Graph

In the case of each function, once the general form of the graph has been identified, the specifics of the function such as the roots, y-intercept and other characteristics can be used to determine the equation that best fits the graph.

Lines

For instance, with graphs of lines, by calculating (or estimating) the slope of the line (m) and the y-intercept of the line (b), the equation for the line can be determined exactly (or approximately, if estimating). In the above case, it is clear that the slope is about 2 and the y-intercept is about 0, thus leading to the equation $y = 2x$.

Step functions

The same type of approach can be used for step functions. First, identify the location of the step and determine the value of the function on either side. In the simplest case, the step function is simply a number on either side, but more complicated cases may involve functions on either side. In any case, once the value or function on either side is determined, identify which side includes the y value (if any) at the x value where the step is located. This allows reconstruction of at least a good approximation of the equation that defines the graph.

Absolute value functions

The slope discontinuity (bend) of an absolute value function is the point at which the expression inside the absolute value brackets is zero. Thus, in the case of the above example, it is clear that the expression in the absolute value term must be $x - 0.5$. Nevertheless, the value of the function at that point is not 0, so there must be an additional term outside the absolute value. In the above case, it turns out to be 0.5. Since the slope of the line is clearly 1 or -1, the absolute value term has a coefficient of 1.

Parabolas

The equation corresponding to the graph of a parabola can be found by looking at the roots, if they exist. If there are two roots, as with the example above, then the equation can be found easily. Assume that the roots are α and β; the equation for the parabola is then the following:

$$y = \gamma(x - \alpha)(x - \beta) = \gamma x^2 - \gamma (\alpha + \beta) x + \gamma \alpha \beta$$

Here, γ is a proportionality factor. Applying this to the example parabola given above, the function is

$$y = \gamma(x - 0)(x - 0.5) = \gamma x^2 - \frac{\gamma}{2} x$$

To find γ, choose a value for x where the function is clearly defined. For instance, look at the function at $x = 1$: $y = 1$ as well. Then:

$$y = 1 = \gamma(1)^2 - \frac{\gamma}{2} (1) = \frac{\gamma}{2}$$

Then $\gamma = 2$. Thus, the equation of the parabola in that case is $y = 2x^2 - x$, as expected.

Exponentials

When determining the equation of a simple exponential function, look first at the y-intercept. Since, for a simple exponential function, the y-intercept is at $x = 0$, this intercept is also the value of a in the equation $y = ae^{bx}$. To determine b, look at the slope of the curve: if the slope is negative, then b is negative; if the slope is positive, then b is positive. The number b can be estimated by finding the value of x where the function is equal to either ae or $\frac{a}{e}$ (e is equal to approximately 2.72). This value of x is equal to either $\frac{1}{b}$ or $\frac{-1}{b}$. For instance, in the example exponential graph above (where $a = 1$), the function is equal to about 0.35 or 0.4 at $x = -1$, which means that b is equal to about 1 (since $e^{(-1)(1)} = \frac{1}{e} = 0.37$). Thus, the equation of the example graph is (approximately) $y = e^x$.

Other functions

For other functions, the approach to determining the corresponding equation for a particular graph follows a similar pattern. By using the characteristics of the function (including roots, the y-intercept and the function value at certain other x values), the parameters of an equation can be determined once it is discovered that the function has a certain type.

The Relationship Between Equations and Graphs

The preceding discussion illustrates methods for deriving information about simple equations using graphical plots. In the other direction, making a graph of an equation is as simple as calculating representative ordered pairs (corresponding x and y values), plotting these points and connecting them appropriately. The greater the number of points plotted, the more accurate the graphical representation.

In light of the fact that graphs are simply a way to depict an equation numerically, it is apparent that the methods described above for determining the equation from the graph are simply a reversal of this logic. The numerical data in the graph is used to reconstruct an equation that is at least approximately representative of the plot.

Thus, if an equation is given, the use of a reverse approach (as compared to the one presented above for determining the equation based on the graph) can be used to determine whether a particular graph represents the equation, or the graph can be constructed independently.

> **SKILL 3.4** **Determine the graphical properties and sketch a graph of a linear, step, absolute-value, quadratic, or exponential function**

See also Skill 1.22

By understanding the basic properties of linear, step, absolute-value, quadratic and exponential functions, it is possible to quickly sketch graphs of these functions. The first step is to determine the limits of the domain that will be plotted on the graph. For instance, to plot the function $y = x$ for x ranging from –1 to 1, it is sufficient to draw axes with tick marks that range (for both the x- and y-axes) from –1 to 1. After drawing the axes, the tick marks should be made at regularly spaced intervals.

The next step in sketching a graph is to select and plot several representative points. The more points that are chosen, the more accurately the graph will represent the function. The points can then be connected with a line to show the curve. It is helpful, whenever possible to use easily determined points such as x- or *y*-intercepts for sketching graphs. This allows sketching reasonably accurate graphs without requiring time-consuming calculations.

Graphing Linear Functions

To graph a line of the form $y = mx + b$, note that b represents the y-intercept (the *y* value for $x = 0$) and *m* represents the slope of the line also referred to as "rise over run" $\left(\dfrac{rise \text{ is the y or vertical distance}}{run \text{ is the } x \text{ or horizontal distance}}\right)$, given by the formula $\dfrac{y_2 - y_1}{x_2 - x_1}$, where (x_1, y_1) and (x_2, y_2) are points on the line. Thus, a graph can be drawn by first plotting the point $(0, b)$ and another convenient point using the slope *m*. For instance, it is possible to also plot $(1, m + b)$.

Example: Plot the line y = 2x − 3.

This line can be plotted by connecting the points (0, –3) and (1, –1) on a set of axes. First, plot the two points and then connect them using a straight edge.

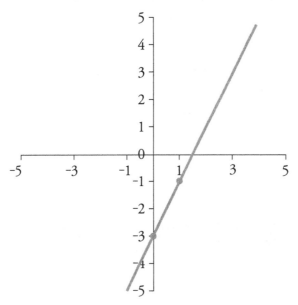

Graphing Step Functions

Step functions can be as simple as a single step (such as the unit step function), where the function has one value on one side of the step and a different value on the other side of the step, or it can involve multiple steps (one situation is the staircase).

Step functions can be represented in a number of ways. The unit step function, for instance, is written as $u(x)$, where *u* is 1 for $x \geq 0$ and *u* is 0 for $x < 0$.

Alternatively, a step function might be written in a piecewise manner, such as the following:

$$f(x) = \begin{cases} a & x > c \\ b & x \le c \end{cases}$$

Staircase step functions might be written as $f(x) = j[rx - h] + k$ or $y = j[rx - h] + k$, where (h, k) is the location of the left endpoint of one step, j is the vertical jump from step to step and r is the reciprocal of the length of each step.

Step functions typically involve rays or line segments with ends that may or may not include a point at the step. This principle is best illustrated by way of examples.

Example: Plot the function

$$f(x) = \begin{cases} 0.5 & x < 1 \\ 1 & 1 \le x \le 2 \\ 0.5 & x > 2 \end{cases}$$

Note that the function is 0.5 everywhere except between 1 and 2 (inclusive). Thus, the line segment that defines the function between these points also includes the values at $x = 1$ and $x = 2$. The rays on either side of this segment do not include a function value at $x = 1$ or $x = 2$. (Note that the definition of the function between $x = 1$ and $x = 2$ uses less (greater) than or equal to notation, but the rays use exclusively less (greater) than notation.)

To sketch the graph, draw the lines and use solid points at the ends to indicate where the function is defined, but use open points to indicate where the function is not defined.

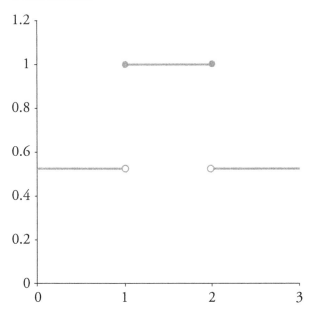

Graphing Absolute Value Functions

The ABSOLUTE VALUE FUNCTION for a first degree equation is of the form $y = m\,|x - h| + k$. The graph is in the shape of a $\vee$ with the point (h, k) being the location of the maximum (or minimum) point on the graph. The sign of m indicates whether the point (h, k) is a maximum or minimum; if m is positive, the graph has a minimum, and, if m is negative, the graph has a maximum. The magnitude of m also determines the slope of the ray on either side of the maximum or minimum point. Consider the following examples.

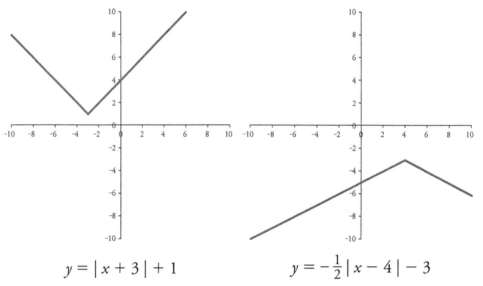

$$y = |x + 3| + 1 \qquad\qquad y = -\frac{1}{2}\,|x - 4| - 3$$

To sketch a graph of an absolute value function, first either convert the function into the form $y = m\,|x - h| + k$ or simply use the function in its current form. Then plot additional points as necessary to determine the location of the rays. Finally, connect the points with a straight edge.

Example: Plot the function $y = |x - 1| + 1$.
Note that this function is already in the form of $y = m\,|x - h| + k$, and the location of (h, k) is simply $(1, 1)$. The slopes of the rays on either side of this point are 1 and -1. Thus, possible other points for plotting include $(2, 2)$ and $(0, 2)$.

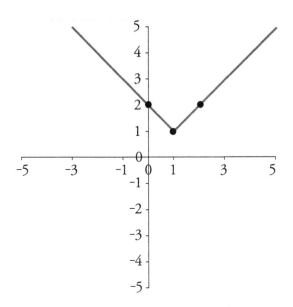

Graphing Quadratic Functions

Because quadratic functions do not involve straight lines, sketching their graphs requires slightly more freehanded technique than does sketching the graphs of linear, step, or absolute value functions. The general form of a quadratic equation is $y = ax^2 + bx + c$. In this form, the y-intercept (at point $(0, c)$) is apparent, along with the direction of concavity for the parabola (upward if a is positive and downward if a is negative). The minimum (or maximum) value of the quadratic function is located at $x = -\frac{b}{2a}$.

If the quadratic function can be factored, then the roots of the equation can be found easily. These roots are the x-intercepts of the function. Alternatively, the quadratic formula can be used to determine the roots, although this can be a more time-consuming method.

Example: Sketch the graph of the function $y = 2x^2 - x + 1$.
Note that the y-intercept for this function is at $(0, 1)$ and that the function is concave up (since the quadratic term is positive). Also note that the minimum value of the function is at the point $(\frac{1}{4}, \frac{7}{8})$. To further aid in sketching the graph, calculate two other points, one on either side of the function minimum. For instance, choose $x = -1$ and $x = 2$. These two values correspond to the points $(-1, 4)$ and $(2, 7)$, respectively. After plotting these points, sketch the curve between them, keeping in mind the general shape of a parabola.

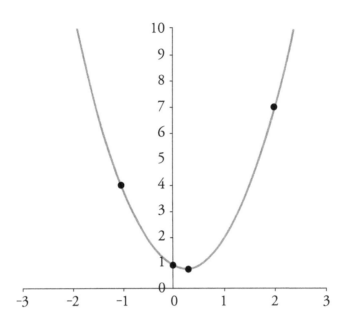

Example: Sketch the graph of the function y = x² − x − 2.

This function can be factored: $y = (x - 2)(x + 1)$, indicating that the roots are at $x = -1$ and $x = 2$. Furthermore, the y-intercept of the parabola is at $(0, -2)$. The minimum point is at $x = \frac{1}{2}$, corresponding to the point $(\frac{1}{2}, \frac{9}{4})$. Plot these points and sketch the curve as shown below.

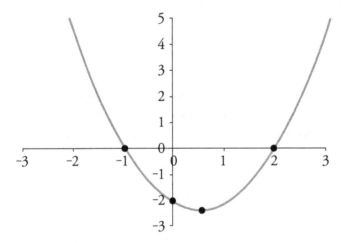

Graphing Exponential Functions

Simple exponential functions have the form $y = ae^{bx}$. The y-intercept of an exponential of this form is always at $y = a$. If a is positive, then the function is entirely positive; if a is negative, the function is entirely negative. Furthermore, if a and b are either both positive or both negative, the slope of the function is positive. If one is negative and the other positive, then the slope is negative. As with the other functions, sketching a graph of an exponential can be done easily by plotting several points.

In addition, it is noteworthy that the function goes to either positive or negative infinity (or to zero) asymptotically as x gets very large (in the negative or positive sense), all depending on the values of a and b.

Example: Sketch a graph of the function $y = -3e^{2x}$.
The y-intercept of this function is $(0, -3)$. Also, the function is negative and has a negative slope. As x gets large in the positive direction, the function approaches negative infinity, and, as x gets large in the negative direction, the function approaches zero. Note that the points corresponding to $x = \frac{1}{2}$ and $x = -\frac{1}{2}$ are $(\frac{1}{2}, -3e)$ and $(-\frac{1}{2}, -\frac{3}{e})$, respectively. The value of e is approximately 2.72, so the decimal values of the points are approximately $(0.5, -8.15)$ and $(-0.5, -1.10)$.

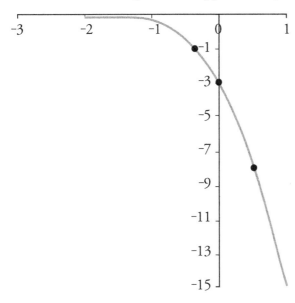

Demonstrate an understanding of a physical situation or a verbal description of a situation and develop a model of it, such as a chart, graph, equation, story, or table

The type of model used to describe a situation depends largely on the details of the situation. A model is designed primarily to organize the information in a situation such that either the data presented can be described in a coherent manner or such that reasonable predictions can be made. It is helpful, when analyzing a situation, to determine if the situation involves a continuous or discrete set of numbers and what form those numbers take. For instance, if a situation involves money in the form of dollars, then the numbers involved are real numbers in increments of $0.01 (one cent). For each situation, determine how the data for a model should best be displayed.

Example: Develop a model of a family budget that allocates 35% to housing, 10% to food, 20% to travel, 25% to taxes, and the remainder to miscellaneous items.

This situation might best be modeled with a pie chart. Note that the miscellaneous category includes 10% of the total budget.

Family Budget

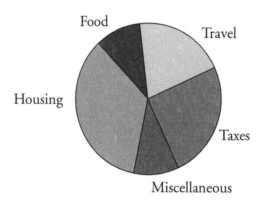

Example: Describe mathematically the relationship between Celsius and Fahrenheit temperature scales (note that the freezing point of water is 0°C and 32°F, and that the boiling point of water is 100°C and 212°F).

In this case, an equation for converting between Fahrenheit and Celsius would be one possible way to model the situation. Let the function $F(C)$ be the conversion function for a temperature in Celsius scale (C) to a temperature in Fahrenheit scale (F). Note that $F(0) = 32$ and $F(100) = 212$. Assuming the relationship between the two scales is linear, the general equation would be the following:

$F(C) = mC + b$

where m and b must be determined from the data given. First, solve for b by using F(0):

$F(0) = 32 = b$

Next, solve for m.

$F(100) = m(100) + 32 = 212$

$212 - 32 = 180 = 100m$

$m = \frac{180}{100} = \frac{9}{5}$

Thus, the result is an equation that mathematically describes the relationship between the Fahrenheit and Celsius temperature scales:

$$F(C) = \tfrac{9}{5}C + 32$$

SKILL **Determine whether a particular mathematical model, such as**
3.6 **an equation, can be used to describe two seemingly different**
situations *(e.g., given two different word problems, determine whether a particular equation can represent the relationship between the variables in the problems).*

Numbers are an abstract representation of objects that allow counting, adding, subtracting, and numerous other operations without reference to specific objects. For instance, three apples and three oranges have a quantity in common; thus, this quantity is abstracted to a number (three in this case). In the same way, equations can be abstracted from particular scenarios, thus allowing the possibility that certain equations may apply to entirely different situations. For instance, the speed of a car and the profile of a ramp may have the same general function (ignoring units).

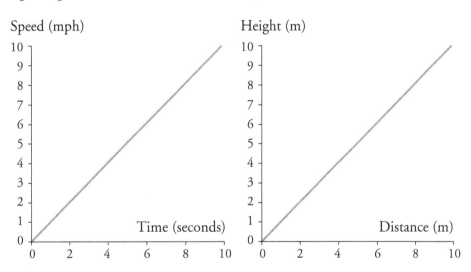

In both cases above, the function could be represented as $y = x$, assuming the horizontal axis is the x-axis and the vertical axis is the y-axis. Nevertheless, although these two graphs can both be described with the same function, they represent two entirely different situations. In the first case, the speed of an accelerating car is shown as a function of time; in the second case, the height of a ramp is shown as a function of distance. The parameters and units of the two situations are entirely different, but, again, they can both be represented by the same equation.

Throughout mathematics and physical sciences, identical equations are often used to describe vastly different phenomena. Since, in such a case, it may be helpful to use the results of one analysis in an analysis of another problem that involves the same equation or set of equations, it is important to be able to identify cases where dissimilar situations permit representation with identical equations.

One of the keys to comparing the representations of two different situations is abstraction. The units used in one situation are often different from those used in another, so they must be ignored. In addition, the types of symbols used may also differ from case to case, so consistent replacement of symbols in particular equations may be necessary.

Example: Determine if the following two equations are the same:
  $$f(\theta) = \sin[d^2\theta], \, y(x) = \sin[\alpha^2 x].$$

At first glance, the equations look different due to differences in the symbols used. If the symbols are renamed consistently, however, the fact that the equations are identical becomes apparent.

Let $y \rightarrow f$, $x \rightarrow \theta$ and $\alpha \rightarrow d$. By applying these replacements to the second equation above, the first equation is obtained. Thus, the two equations are the same (although, technically, d and α may differ between two situations). If the functions were $f(\theta) = \sin[\theta]$ and $y(x) = \sin[x]$, then they would be unambiguously identical.

The process of determining whether a particular mathematical model, such as an equation, applies identically to two different situations therefore requires deriving a model for one situation and comparing it to a derived model for another situation. Consider the following examples.

Example: Determine whether the following two situations can be modeled with the same type of mathematical equation—for example, the position of a billiard ball in uniform motion as it bounces off a bumper on a pool table or the height of a peaked roof on a house.
Consider the motion of a billiard ball as it strikes a bumper and then bounces off. This motion looks approximately like the following:

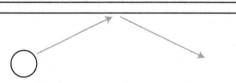

An equation could be constructed to show the distance of the ball from the bumper with respect to time. This equation would involve an absolute value function because of the discontinuity of motion when the ball hits the bumper.

In the case of a peaked house roof, the profile of the roof might look like the following:

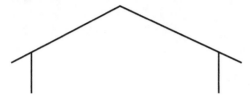

An absolute value equation could likewise be used here to express the height of the roof as a function of distance from one side of the house. Depending on the specific details of each situation, the numbers involved in the equations may differ, but the basic form of the equations will be the same.

Example: Determine if the following two word problems can be solved using the same mathematical model:

1. Write the equation for the position (in miles) of a car traveling in a straight line at a speed v (in miles per hour).

2. Write the equation for the depth of water (in feet) in a circular pool of radius r (in feet) being filled at a rate of f (in cubic feet per hour).

Notice that, although the parameters of the problem are different, both problems involve constant rates. Thus, it can generally be expected that both problems will involve linear equations of some sort.

For the first problem, the position of the car after some time t is simply the product of the speed, v, and t. Therefore, if we define position s in miles and t in hours,

$$s = vt$$

For the second problem, the volume of water in the pool is the area of the pool multiplied by the height of the water. Additionally, the water volume is also equal to the product of the rate of water flowing into the pool and the elapsed time t in hours. This leads to the following equation, which can be solved for the depth d in feet:

$$\pi r^2 d = ft$$
$$d = \frac{f}{\pi r^2} t$$

Clearly the two equations are similar, but if it so happens that $\frac{f}{\pi r^2}$ is equal to v, then the two equations would be identical (apart from the defined units). Thus, in this case, the two word problems involve the same mathematical model, or, in other words, the same equation can be used to represent the relationship between the variables of both problems.

SKILL **Find the domain (*x*-values) and range (*y*-values) of a function**
3.7 **without necessarily knowing the definitions; recognize certain**
properties of graphs *(e.g., slope, intercepts, intervals of increase or decrease,*
axis of symmetry)

The Domain and Range of a Function

DOMAIN: the set of all (*x*) values for which the function is defined

RANGE: the set of all (*y*) values that correspond, by way of the function, to the values in the domain

Generally, the DOMAIN of a function is the set of all (*x*) values for which the function is defined. (Recall that a function assigns a single *y* value to each *x* value in the domain.) The RANGE of a function is the set of all (*y*) values that correspond, by way of the function, to the values in the domain.

For typical algebraic functions (of the form *y(x)*), the domain can be found by determining what set of numbers corresponds to a single defined output of the function. By looking at a graph, for instance, one can note whether there are points for which the function does not have a defined value. In such a case, there may be a missing point (depicted as an open circle in the plot) or there may be a vertical asymptote indicating that the function approaches positive or negative infinity as it approaches a certain value. Consider the following examples.

Example: Determine the domain of the function depicted in the following graph.

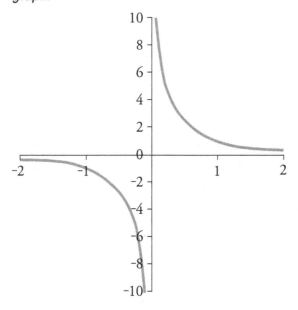

This function has two asymptotes: one for *y* = 0 and one for *x* = 0. It is apparent that the function is not defined for *x* = 0, but that it has finite values everywhere else. Thus, the domain of the function is all real numbers except 0.

The function plotted here is $y = \frac{1}{x}$; thus, by way of the function, it is clear that the domain includes all real values except 0, for which the function goes to either positive or negative infinity in the limit (depending on the direction).

Example: Determine the domain of the function in the graph.

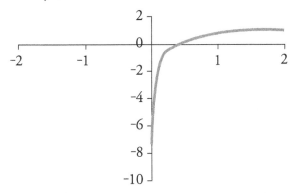

In this case, the function is clearly not defined for any value less than zero (the function is entirely absent). Furthermore, there is an asymptote at $x = 0$, thus indicating that the function is not defined for $x = 0$ as well. Therefore, the domain of the function is all positive real numbers.

The actual function depicted here is $x = \ln x$. This function is not defined for $x = 0$ or for negative values of x. Thus, the results obtained for the graph are shown to be correct.

Properties of Graphs

Slope

Identification of the properties of a graph can be helpful in determining the corresponding function that is being plotted (either exactly or approximately), or can provide information about the behavior of functions over certain intervals. For instance, the SLOPE of the graph is the rate of increase in the vertical direction for a given increase in the horizontal direction. The slope is defined as positive when the plot of the function increases when going from left to right, and it is defined as negative when the function decreases when going from left to right.

> **SLOPE:** the rate of increase in the vertical direction for a given increase in the horizontal direction

Only lines have a constant slope; other functions generally have a slope that varies continuously over the domain of the function. For instance, the function shown below has a varying slope that is positive for positive x values and negative for negative x values.

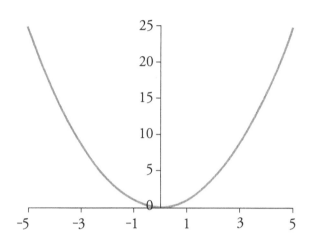

Intervals of increase and decrease

The INTERVALS OF INCREASE OR DECREASE for a function are those respective portions of the domain for which the function is increasing monotonically or decreasing monotonically. In the example graph above, the function has an interval of decrease over $(-\infty,0)$ and an interval of increase over $(0,\infty)$.

Axis of symmetry

Furthermore, the function graphed above has an AXIS OF SYMMETRY at $x = 0$. An axis of symmetry is simply a line that divides two symmetric portions of a function. The function displays a mirror image across such a line, with the axis of symmetry acting as the hypothetical mirror.

Intercepts

The INTERCEPTS of a function are those points at which the function crosses one or both of the axes. Since a function has only one value in the range corresponding to each value in the domain, a function can have only one y-intercept. Nevertheless, a function can have infinitely many x-intercepts. Consider the graph of the cosine function, as shown below.

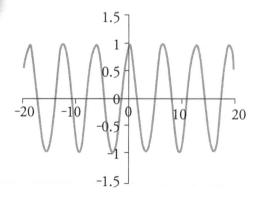

**INTERVALS OF IN-
CREASE OR DECREASE:**
those respective portions
of the domain for which
the function is increasing
monotonically or decreas-
ing monotonically

AXIS OF SYMMETRY:
a line that divides two
symmetric portions of a
function

INTERCEPTS: those
points at which the func-
tion crosses one or both of
the axes

The cosine function has a y-intercept at $y = 1$ and numerous x-intercepts (the total number of x-intercepts is infinite, but only 12 are shown here). The x-intercepts are at $x = \frac{n\pi}{2}$, where $n = \pm1, \pm3, \pm5$ and so on.

Example: Identify the critical properties of the following graph.

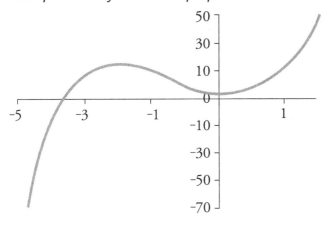

The function depicted above has no apparent axes of symmetry. It has a y-intercept at about $y = 2$ and an x-intercept at about $x = -3.7$. The intervals of increase are (extrapolating from the given information) approximately $(-\infty, -2.6)$ and $(0, \infty)$. If only the information in the visible graph is used, and no assumptions about the function outside the limits of the graph are made, then the intervals of increase are about $(-5, -2.6)$ and $(0, 2)$. The lone interval of decrease is about $(-2.6, 0)$.

The slope of the graph can be approximated, if necessary, on a point by point basis. It is noteworthy that the slope of the graph is zero at about $x = -2.6$ and $x = 0$.

SKILL **Translate verbal expressions and relationships into algebraic**
3.8 **expressions or equations; provide and interpret geometric**
representations of numeric and algebraic concepts

Example: Write the following statement as an algebraic equation: "The height of the rocket is the product of the speed and the amount of time in flight, plus the starting height."

First, define the variables involved in the problem. Express the height of the rocket, h, in terms of the speed; v, the time in flight, t; and the initial height, s. If ambiguity arises, such as might take place were the statement spoken instead of written (the lack of a comma might lead to the question as to whether the second term in the product is t or $(t + s)$), choose the interpretation that makes the most sense according to the situation.

Next, write the equation using the defined variables:

$$h = vt + s$$

This equation, along with the definitions of each parameter, constitutes the solution to this problem.

Example: Write an equation based on the following statement: "I want to know the price of a single banana if bananas are sold by the dozen for x dollars."

In this case, what is needed is the given price, x, divided by 12, since this gives the price per banana, b. This is written in algebraic form as follows:

$$b = \frac{x}{12}$$

The preceding examples are indeed simple, but they are representative of the types of processes needed to convert verbal or written statements into algebraic expressions. Mastery of this skill is a significant portion of solving word problems, for instance.

Geometric Representations of Numeric and Algebraic Concepts

Geometry can be a useful tool in explaining or further understanding numeric and algebraic concepts. For instance, the principal of commutativity in multiplication can be illustrated with a rectangle. The area of a rectangle is the product of the length and the width.

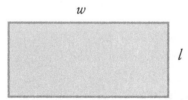

Note, however, that if we redefine the length and the width (that is, reverse the definitions), the area of the rectangle remains the same. Thus, the following two cases are equivalent.

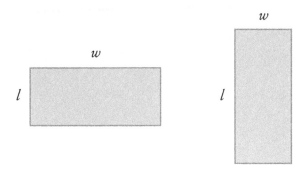

As a result, $l \times w = w \times l = Area$.

Likewise, other geometric interpretations can be derived for various numeric and algebraic concepts. Addition, for example, can be represented as the linking of line segments of given lengths relative to a number line.

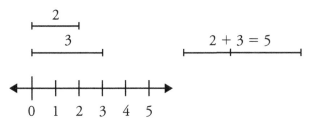

Example: Provide a geometric interpretation of the following inequality:
$|a| + |b| \geq |a + b|$.

Let a and b be line segments, and $|a|$ is the length of a, $|b|$ is the length of b, and $|a + b|$ is the length between the endpoints of the segments a and b, however joined (they may, for instance, be joined at an angle, as shown following).

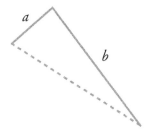

Notice that the length of the dashed line can only be as great as the sum of the lengths of a and b. Therefore, the inequality can be interpreted in terms of triangles if $|a + b|$ is the length of the dashed line. (This example can be illustrated more clearly with vectors, but the use of line segments in this manner is sufficient.)

DOMAIN IV
DATA, PROBABILITY, AND STATISTICAL CONCEPTS AND DISCRETE MATHEMATICS

PERSONALIZED STUDY PLAN

✗ KNOWN MATERIAL/ SKIP IT

PAGE	COMPETENCY AND SKILL	
187	**4A: Data, probability, and statistical concepts**	☐
	4A.1: Organize data into a presentation that is appropriate for solving a problem	☐
	4A.2 Read and analyze data presented in various forms	☐
	4A.3 Solve probability problems involving finite sample spaces using counting techniques	☐
	4A.4 Solve probability problems involving independent and dependent events	☐
	4A.5 Solve problems by using geometric probability	☐
	4A.6 Solve problems involving average	☐
	4A.7 Find and interpret common measures of central tendency	☐
	4A.8 Find and interpret common measures of dispersion	☐
207	**4B: Discrete mathematics**	☐
	4B.1 Use and interpret statements that contain logical connectives and logical quantifiers	☐
	4B.2 Solve problems involving the union and intersection of sets, subsets, and disjoint sets	☐
	4B.3 Solve basic counting problems involving permutations and combinations; use Pascal's triangle to solve problems	☐
	4B.4 Solve problems that involve simple sequences or number patterns	☐
	4B.5 Use and interpret matrices as tools for displaying data	☐
	4B.6 Draw conclusions from information contained in simple diagrams, flowcharts, paths, circuits, networks, or algorithms	☐
	4B.7 Explore patterns in order to make conjectures, predictions, or generalizations	☐

✗

COMPETENCY 4A

DATA, PROBABILITY, AND STATISTICAL CONCEPTS

Construction of histograms is discussed in Skill 4A.2

We can use a table, graph, or rule to show a relationship between two quantities. In the following example, the rule $y = 9x$ describes the relationship between the total amount earned, y, and the total amount of $9 sunglasses sold, x.

A table using this data would appear as follows:

Domain IV has two subsections, each with a set of skills: A—Data, Probability, and Statistical Concepts, and B—Discrete Mathematics.

Number of Sunglasses Sold	1	5	10	15
Total Dollars Earned	9	45	90	135

To read a chart or table, read the row and column headings. Use this information to evaluate the given information in the chart or table.

Each (x, y) relationship between a pair of values is called a coordinate pair and can be plotted on a graph. The coordinate pairs (1, 9), (5, 45), (10, 90), and (15, 135) are plotted on the following graph.

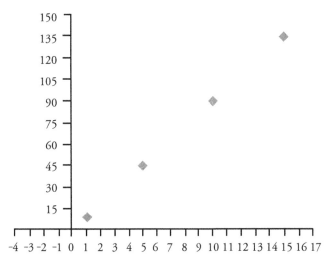

The graph shows a linear relationship. A linear relationship is one in which two quantities are proportional to each other. Doubling x also doubles y. On a graph, a straight line depicts a linear relationship.

We can analyze the function or relationship between two quantities to determine how one quantity depends on the other. For example, the following function shows a relationship between y and x:

$y = 2x + 1$

The function $y = 2x + 1$ is written as a symbolic rule. The following table shows the same relationship:

x	0	2	3	6	9
y	1	5	7	13	19

We can write a relationship in words by saying the value of y is equal to two times the value of x, plus one. We can show this relationship on a graph by plotting given points such as those shown in the table above.

SKILL 4A.2 **Read and analyze data presented in various forms** (e.g., tables, charts, graphs, line, bar, histogram, circle, double line, double bar, scatterplot, stem and plot, line plot, box plot); draw conclusions from data

Tables

Example: The results of a survey of 47 students are summarized in the table below.

	BLACK HAIR	BLONDE HAIR	RED HAIR	TOTAL
Male	10	8	6	24
Female	6	12	5	23
Total	16	20	11	47

Use the table to answer questions a–c.

A. If one student is selected at random, find the probability of selecting a male student

$$\frac{\text{Number of male students}}{\text{Number of students}} = \frac{24}{47}$$

B. If one student is selected at random, find the probability of selecting a female with red hair

$$\frac{\text{Number of red hair females}}{\text{Number of students}} = \frac{5}{47}$$

C. If one student is selected at random, find the probability of selecting a student that does not have red hair

$$\frac{\text{Red hair students}}{\text{Number of students}} = \frac{11}{47}$$

$$1 - \frac{11}{47} = \frac{36}{47}$$

Line Graphs

Line graphs are used to show trends, often over a period of time. To make a line graph, determine appropriate scales for both the vertical and horizontal axes (based on the information to be graphed). Describe what each axis represents and mark the scale periodically on each axis. Graph the individual points of the graph and connect the points on the graph from left to right.

	TEST 1	TEST 2	TEST 3	TEST 4	TEST 5
Evans, Tim	75	66	80	85	97
Miller, Julie	94	93	88	97	98
Thomas, Randy	81	86	88	87	90

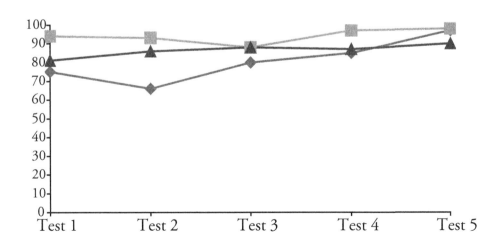

Example: Graph the following information using a line graph.

THE NUMBER OF NATIONAL MERIT FINALISTS PER SCHOOL YEAR						
YEAR	90–91	91–92	92–93	93–94	94–95	95–96
Central	3	5	1	4	6	8
Wilson	4	2	3	2	3	2

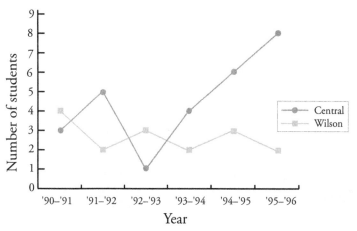

Bar Graphs

BAR GRAPHS are used to compare various quantities. To make a bar graph or a pictograph, determine the scale to be used for the graph. Then determine the length of each bar on the graph or determine the number of pictures needed to represent each item of information. Be sure to include an explanation of the scale in the legend if the numbers on the axes do not represent the actual numbers.

	Test 1	Test 2	Test 3	Test 4	Test 5
Evans, Tim	75	66	80	85	97
Miller, Julie	94	93	88	97	98
Thomas, Randy	81	86	88	87	90

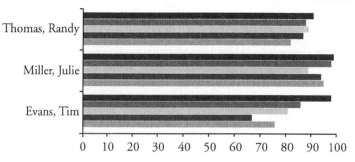

Example: A class had the following grades: 4 A's, 9 B's, 8 C's, 1 D, 3 F's. Graph these on a bar graph.

Bar graph

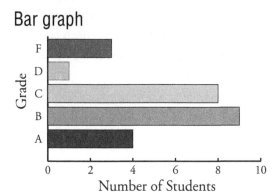

To read a bar graph, read the explanation of the scale that was used in the legend (if the data has been scaled). Compare the length of each bar with the dimensions on the axes and calculate the value each bar represents.

Histograms

A HISTOGRAM is used to summarize the information in a large data set. It shows the counts of data in different ranges. The count in each range or bin is known as the frequency. The histogram shows graphically the center of the data set, the spread of the data and whether there are any outliers. It also shows whether the data has a single mode or more than one.

> **HISTOGRAM:** used to summarize the information in a large data set

Example: The histogram below shows the summary of some test results where people scored points ranging from 0 to 45. The total range of points has been divided into bins 0–5, 6–10, 11–15, and so on. The frequency for the first bin (labeled 5) is the number of people who scored points ranging from 0 to 5; the frequency for the second bin (labeled 10) is the number of people who scored points ranging from 6 to 10 and so on.

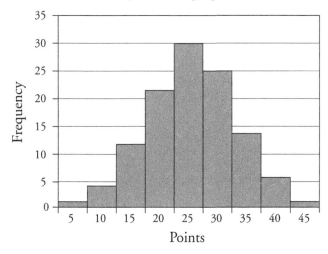

> **DISCRETE FREQUENCY DISTRIBUTION:** used to represent discrete as well as continuous data (data that can take on a continuous range of values—e.g., height) sorted in bins

A histogram is a DISCRETE FREQUENCY DISTRIBUTION. It can be used to represent discrete as well as continuous data (data that can take on a continuous range of values—e.g., height) sorted in bins. A large data set of continuous data may also be represented using a continuous frequency distribution, which is essentially a histogram with very narrow bars. Below, a trend line has been added to the example histogram above. Notice that this approximates the most common continuous distribution, a normal or bell curve.

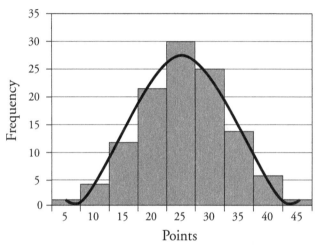

A NORMAL DISTRIBUTION is a symmetric distribution, with the mean equal to the median. The tails of the curve in both directions fall off rapidly. The spread of data is measured by the standard deviation.

> **NORMAL DISTRIBUTION:** a symmetric distribution, with the mean equal to the median

> *For more information on frequency distrubutions see:*
>
> *http://mathworld.wolfram. com/ContinuousDistribu- tion.html*

Circle Graphs

CIRCLE GRAPHS show the relationship of various parts to each other and to the whole. Percents are used to create circle graphs.

Example: Julie spends 8 hours each day in school, 2 hours doing homework, 1 hour eating dinner, 2 hours watching television,10 hours sleeping, and the rest of the time doing other activities.

> **CIRCLE GRAPHS:** show the relationship of various parts to each other and to the whole

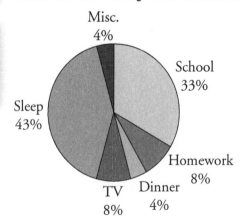

To make a circle graph, first total all of the information that is to be included on the graph. Then determine the central angle to be used for each sector of the graph using the following formula:

$$\frac{\text{information}}{\text{total information}} \times 360° = \text{degrees in central} \measuredangle$$

Lay out the central angles to these sizes, label each section, and include its percent.

Example: Graph this information on a circle graph:

Monthly expenses:

Rent: $400

Food: $150

Utilities: $75

Clothes: $75

Activities: $100

Misc.: $200

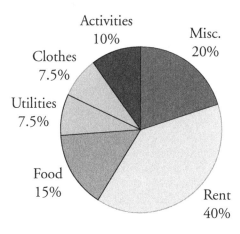

To read a circle graph, find the total of the amounts represented on the entire circle graph. To determine the actual amount that each sector of the graph represents, multiply the percent in a sector times the total amount number.

Scatterplots

SCATTERPLOTS compare two characteristics of the same group of things or people and usually consist of a large body of data. They show how much one variable

> **SCATTERPLOTS:** plots that compare two characteristics of the same group of things or people and usually consist of a large body of data

CORRELATION: the relationship between two variables

affects another. The relationship between the two variables is their CORRELATION. The closer the data points come to making a straight line when plotted, the closer the correlation.

Correlation is a measure of association between two variables.

CORRELATION COEF-FICIENT: describes the strength of the association between the variables and the direction of the association

The CORRELATION COEFFICIENT (r) is used to describe the strength of the association between the variables and the direction of the association. It varies from –1 to 1, with 0 being a random relationship, 1 being a perfect positive linear relationship, and –1 being a perfect negative linear relationship.

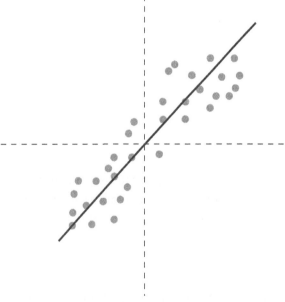

Horizontal and vertical lines are drawn through the point of averages, which is the point on the averages of the x and y values. This divides the scatterplot into four quadrants. If a point is in the lower left quadrant, the product of two negatives is positive; in the upper right, the product of two positives is positive.

In the two remaining quadrants (upper left and lower right), the product of a negative and a positive is negative. If r is positive, then there are more points in the positive quadrants and if r is negative, then there are more points in the two negative quadrants.

Stem and Leaf Plots

STEM AND LEAF PLOTS are visually similar to line plots. The stems are the digits in the greatest place value of the data values, and the leaves are the last digits in the data. Stem and leaf plots are best suited for small sets of data and are especially useful for comparing two sets of data.

The first thing to do when creating a stem and leaf plot is to arrange the data from smallest to largest. The second thing to do is to list the range of scores for the stems in a column. Finally, you take each score, one at a time and put the last digit in a column next to its stem.

The following is an example using test scores
The test scores are 49, 54, 59, 61, 62, 63, 64, 66, 67, 68, 68, 70, 73, 74, 76, 76, 76, 77, 77, 77, 77, 78, 78, 78, 78, 83, 85, 85, 87, 88, 90, 90, 93, 94, 95, 100, and 100.

STEM	LEAVES
4	9
5	4 9
6	1 2 3 4 6 7 8 8
7	0 3 4 6 6 6 7 7 7 8 8 8 8
8	3 5 5 7 8
9	0 0 3 4 5
10	0 0

> **STEM AND LEAF PLOTS:** similar to line plots, these plots are best suited for small sets of data and are especially useful for comparing two sets of data

Line Plots

LINE PLOTS are usually used for one group of data with fewer than 50 values. A line plot consists of a horizontal line with an x marked above a value for each occurrence of that value in the data set.

> **LINE PLOTS:** plots used for one group of data with fewer than 50 values

Example:

Each of 25 workers in a candle shop took a box of 20 candles and counted the number of red candles in the box. The distribution of red candles in their boxes is shown below.

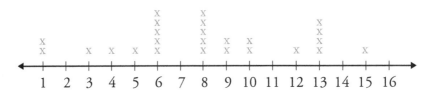

The line plot indicates that 2 workers had 1 red candle, 1 had 3, 1 had 4, 1 had 5, 5 had 6, 5 had 8, 2 had 9, 2 had 10, 1 had 12, 4 had 13, and 1 had 15.

Box-and-Whiskers Plots

BOX-AND-WHISKERS PLOT: plots that display five statistics: a minimum score, a maximum score, and three percentiles

A BOX-AND-WHISKERS PLOT displays five statistics: a minimum score, a maximum score, and three percentiles. A percentile value for a score tells you the percentage of scores lower than it. The beginning of the box is the score at the 25th percentile. The end of the box represents the 75th percentile. The score inside of the box is the median, or the score at the 50th percentile. The score at the end of the left whisker is the minimum score. The score at the end of the right whisker is the maximum score.

Class 1

The above box-and-whiskers plot summarizes the scores on a mathematics test for the students in Class I. It indicates that the lowest score is 60, while the highest score is 100. Twenty-five percent of the class scored 68 or lower, 50% scored more than 76, and 25% scored 90 or higher.

Pictographs

PICTOGRAPH: use simple drawings to represent a set of data

To make a PICTOGRAPH, determine the scale to be used for the graph. Then determine the number of pictures needed to represent each item of information. Be sure to include an explanation of the scale in the legend if each picture represents more than one item of information.

*Example: A class had the following grades: 4 A's, 9 B's, 8 C's, 1 D, 3 F's.
Graph these on a pictograph.*

Grade	Number of Students
A	☺☺☺☺
B	☺☺☺☺☺☺☺☺☺
C	☺☺☺☺☺☺☺☺
D	☺
F	☺☺☺

Solve probability problems involving finite sample spaces by actually counting outcomes; solve probability problems by using counting techniques

Probability and Counting Techniques

In probability, the SAMPLE SPACE is a list of all possible outcomes of an experiment. For example, the sample space of tossing two coins is the set {HH, HT, TT, TH}, where heads is H and tails is T, and the sample space of rolling a six-sided die is the set {1, 2, 3, 4, 5, 6}.

When conducting experiments with a large number of possible outcomes, it is important to determine the size of the sample space. The size of the sample space can be determined by using the fundamental counting principle and the rules of combinations and permutations.

The FUNDAMENTAL COUNTING PRINCIPLE states that if there are m possible outcomes for one task and n possible outcomes of another, there are ($m \times n$) possible outcomes of the two tasks together.

A PERMUTATION is the number of possible arrangements of items, without repetition, where order of selection is important.

A COMBINATION is the number of possible arrangements, without repetition, where order of selection is not important.

SAMPLE SPACE: a list of all possible outcomes of an experiment

FUNDAMENTAL COUNTING PRINCIPLE: states that if there are m possible outcomes for one task and n possible outcomes of another, there are ($m \times n$) possible outcomes of the two tasks together

PERMUTATION: the number of possible arrangements of items, without repetition, where order of selection is important

COMBINATION: the number of possible arrangements, without repetition, where order of selection is not important

Example: If any two numbers are selected from the set {1, 2, 3, 4}, list the possible permutations and combinations.

COMBINATIONS	PERMUTATIONS
12, 13, 14, 23, 24, 34: 6 ways	12, 21, 13, 31, 14, 41, 23, 32, 24, 42, 34, 43: 12 ways

Note that the list of permutations includes 12 and 21 as separate possibilities, since the order of selection is important. In the case of combinations, however, the order of selection is not important and therefore 12 is the same combination as 21. Hence, 21 is not listed separately as a possibility.

The number of permutations and combinations may also be found by using the formulae given below.

The number of possible permutations in selecting r objects from a set of n is given by

$$_nP_r = \frac{n!}{(n-r)!}$$

The notation $_nP_r$ is read "the number of permutations of n objects taken r at a time."

In our example, two objects are being selected from a set of four.

$$_4P_2 = \frac{4!}{(4-2)!}$$ Substitute known values.
$$_4P_2 = 12$$

The number of possible combinations in selecting r objects from a set of n is given by

$$_nC_r = \frac{n!}{(n-r)!r!}$$

The number of combinations when r objects are selected from n objects.

In our example,

$$_4C_2 = \frac{4!}{(4-2)!2!}$$ Substitute known values.
$$_4C_2 = 6$$

Example: Find the size of the sample space of rolling two six-sided dice and flipping two coins.

List the possible outcomes of each event:

    each die: {1, 2, 3, 4, 5, 6}

    each coin: {heads, tails}

Apply the fundamental counting principle:

    size of sample space = $6 \times 6 \times 2 \times 2 = 144$

Example: Find the size of the sample space of selecting 3 playing cards at random from a standard 52-card deck.

Use the rule of combination:

$$_{52}C_3 = \frac{52!}{(52-3)!3!} = 22100$$

> ### SKILL 4A.4   Solve probability problems involving independent and dependent events

Addition Principles of Counting

Counting principles are used to calculate the total number of occurrences or the sample space when there is more than one event. Given two events A and B, the addition principles of counting apply to situations where A *or* B occurs. The multiplication principles of counting apply to situations where A *and* B occur.

In the notation used below, n(A) denotes the number of occurrences or sample space of event A. n(A∩B) denotes the intersection of the sample spaces of A and B.

The Addition Principle of Counting states:

> If *A* and *B* are events, $n(A \text{ or } B) = n(A) + n(B) - n(A \cap B)$

Example: How many ways can you select a black card or a Jack from an ordinary deck of playing cards?

Let *B* denote the set of black cards and let *J* denote the set of Jacks. Then,

$n(B) = 26, \ n(J) = 4, \ n(B \cap J) = 2$

$n(B \text{ or } J) = 26 + 4 - 2 = 28$

The Addition Principle of Counting for Mutually Exclusive Events states:

> If *A* and *B* are mutually exclusive events, $n(A \text{ or } B) = n(A) + n(B)$.

Example: A travel agency offers 40 possible trips: 14 to Asia, 16 to Europe and 10 to South America. How many trips can be taken to either Asia or Europe through this agency?

Let *A* denote trips to Asia and let *E* denote trips to Europe. Then, $A \cap E = \varnothing$ and

$n(A \text{ or } B) = 14 + 16 = 30$.

Therefore, the number of trips to Asia or Europe is 30.

Multiplication Principles of Counting

The Multiplication Principle of Counting for Dependent Events states:

> Let A be a set of outcomes of Stage 1 and B a set of outcomes of Stage 2. The number of ways, $n(A \text{ and } B)$, A and B can occur in a two-stage experiment is given by: $n(A \text{ and } B) = n(A)n(B|A)$, where $n(B|A)$ denotes the number of ways B can occur given that A has already occurred.

Example: How many ways from an ordinary deck of 52 cards can 2 Jacks be drawn in succession if the first card is drawn but not replaced in the deck and then the second card is drawn?

This is a two-stage experiment for which we wish to compute $n(A \text{ and } B)$ where A is the set of outcomes for which a Jack is obtained on the first draw and B is the set of outcomes for which a Jack is obtained on the second draw.

If the first card drawn is a Jack, then there are only three remaining Jacks left to choose from on the second draw. Thus, drawing 2 cards without replacement means the events A and B are dependent.

$$n(A \text{ and } B) = n(A)n(B|A) = 4 \times 3 = 12$$

The Multiplication Principle of Counting for Independent Events states:

> Let A be a set of outcomes of Stage 1 and B a set of outcomes of Stage 2. If A and B are independent events, the number of ways, $n(A \text{ and } B)$, A and B can occur in a two-stage experiment is given by $n(A \text{ and } B) = n(A)n(B)$

Example: How many six-letter combinations can be formed if repetition of letters is not allowed?

A first letter and a second letter and a third letter and a fourth letter and a fifth letter and a sixth letter must be chosen, so there are six stages.

Since repetition is not allowed, there are 26 choices for the first letter, 25 for the second, 24 for the third, 23 for the fourth, 22 for the fifth, and 21 for the sixth
n(six-letter combinations without repetition of letters)

$$= 26 \times 25 \times 24 \times 23 \times 22 \times 21$$
$$= 165{,}765{,}600$$

DEPENDENT EVENTS:
events that occur when the probability of the second event depends on the outcome of the first event

Dependent and Independent Events

DEPENDENT EVENTS occur when the probability of the second event depends on the outcome of the first event. INDEPENDENT EVENTS are unrelated and the probability of the second event is not dependent on the first.

For example, consider the two events. (A) It is sunny on Saturday, and (B) a person will go to the beach. If the person intends to go to the beach on Saturday, rain or shine, then A and B may be independent. If, however, the person plans to go to the beach only if it is sunny, then A and B may be dependent. In this situation, the probability of event B will change depending on the outcome of event A.

> **INDEPENDENT EVENTS:** unrelated events for which the probability of the second event is not dependent on the first

Suppose a pair of dice, one red and one green, are rolled. If a three is rolled on the red die and a four on the green die, the events do not depend on each other. The total probability of the two independent events can be found by multiplying the separate probabilities.

$$P(A \text{ and } B) = P(A) \times P(B)$$
$$= \frac{1}{6} \times \frac{1}{6}$$
$$= \frac{1}{36}$$

Example: A jar contains 12 red marbles and 8 blue marbles. If a marble is randomly selected and replaced, and then a second marble is randomly selected, find the probability of selecting two red marbles.

The selections are independent events because the probability of picking a red marble the second time remains the same.

$$P(\text{Red and Red}) \text{ with replacement} = P(\text{Red}) \times P(\text{Red})$$
$$= \frac{12}{20} \times \frac{12}{20}$$
$$= \frac{9}{25}$$

Example: A jar contains 12 red marbles and 8 blue marbles. If a marble is randomly selected and not replaced, and then a second marble is randomly selected, find the probability of selecting two red marbles.

If you pick a red marble and pick again without replacing the first red marble, the second pick becomes dependent upon the first pick.

$$P(\text{Red and Red}) \text{ without replacement} = P_1(\text{Red}) \times P_2(\text{Red})$$
$$= \frac{12}{20} \times \frac{11}{19}$$
$$= \frac{33}{95}$$

SKILL Solve problems by using geometric probability
4A.5

Geometric probability describes situations that involve shapes and measures. For example, given a 10-inch string, we can determine the probability of cutting the string so that one piece is at least 8 inches long. If the cut occurs in the first or last two inches of the string, one of the pieces will be at least 8 inches long.

String

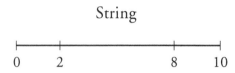

Thus, the probability of such a cut is $\frac{2 + 2}{10} = \frac{4}{10} = \frac{2}{5}$ or 40%.

Other geometric probability problems involve the ratio of areas. For example, to determine the likelihood of randomly hitting a defined area of a dartboard (pictured below) we determine the ratio of the target area to the total area of the board.

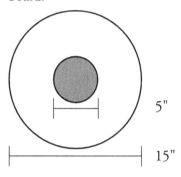

Given that a randomly thrown dart lands somewhere on the board, the probability that it hits the target area is the ratio of the areas of the two circles. Thus, the probability, P, of hitting the target is

$$P = \frac{(2.5)^2\pi}{(7.5)^2\pi} \times 100 = \frac{6.25}{56.25} \times 100 = 11.1\%.$$

SKILL 4A.6 **Solve problems involving average, including arithmetic mean and weighted average**

Expressed symbolically, the arithmetic mean or average is written as:
$$\bar{x} = \frac{1}{n}\sum_{i=1}^{n} X_i$$

The AVERAGE is the sum of n numbers divided by n.

The average of the five numbers 9, 7, 10, 8, 6 is
$$\frac{9 + 7 + 10 + 8 + 6}{5} = 8$$

The weighted average is expressed symbolically as:
$$WM = \sum_{i=1}^{n}\frac{w_i}{w} X_i$$

> **AVERAGE:** the sum of n numbers divided by n

A WEIGHTED AVERAGE involves computing an average where each value to be averaged has an assigned weight. Teachers often compute grades using a weighted average.

Example: Mrs. Wu assigns a weight of 10% to homework, 20% to quizzes, and 70% to tests. Suppose Student A has an average homework grade of 90, an average quiz grade of 92, and an average test grade of 85. Mrs. Wu would compute her total grade in the following manner:

$(0.10)(90) + (0.20)(92) + (0.70)(85) =$

$9 + 18.4 + 59.5 = 86.9$

> **WEIGHTED AVERAGE:** involves computing an average where each value to be averaged has an assigned weight

---

SKILL **Find and interpret common measures of central tendency** *(e.g., mean,*
4A.7 *sample mean, median, mode)* **and know which is the most meaningful to use in a given situation**

Mean, median, and mode

These are three measures of central tendency. The MEAN is the average of the data items. The MEDIAN is found by putting the data items in order from smallest to largest and selecting the item in the middle (or the average of the two items in the middle). The MODE is the most frequently occurring item.

> **MEAN:** the average of the data items

> **MEDIAN:** found by putting the data items in order from smallest to largest and selecting the item in the middle

> **MODE:** the most frequently occurring item

Example: Find the mean, median, mode, and range of the test score listed below:

85	77	65
92	90	54
88	85	70
75	80	69
85	88	60
72	74	95

Mean (X) = sum of all scores × number of scores = 78

Median = put numbers in order from smallest to largest. Pick middle number.

54 60 65 69 70 72 74 75 |77 80| 85 85 85 88 88 90 92 95

both in middle

Therefore, the median is the average of the two numbers in the middle, or 78.5

Mode = most frequent number = 85

Range = largest number minus the smallest number

= 95 − 54

= 41

Different situations require different information. If we examine the circumstances under which the owner of an ice cream store may use statistics collected in the store, we find different uses for different information.

Over a seven-day period, the owner collected data on the ice cream flavors sold. He found that the mean number of scoops sold was 174 per day. The most frequently sold flavor was vanilla. This information was useful in determining how much ice cream to order and in what amounts for each flavor.

In the case of the ice cream store, the median and range had little business value for the owner.

Consider the set of test scores from a math class: 0, 16, 19, 65, 65, 65, 68, 69, 70, 72, 73, 73, 75, 78, 80, 85, 88, and 92. The mean is 64.06 and the median is 71. Since there are only three scores less than the mean out of the 18 scores, the median (71) would be a more descriptive score.

Retail clothing store owners may be most concerned with the most common dress size, so they may order more of that size than any other size.

An understanding of the definitions is important in determining the validity and uses of statistical data. All definitions and applications in this section apply to ungrouped data.

| SKILL 4A.8 | **Find and interpret common measures of dispersion** (e.g., range, spread of data, standard deviation, outliers) |

Range, Percentiles, Stanines, and Quartiles

RANGE: a measure of variability found by subtracting the smallest value from the largest value

RANGE is a measure of variability. It is found by subtracting the smallest value from the largest value.

PERCENTILES divide data into 100 equal parts. A person whose score falls in the 65th percentile has outperformed 65 percent of all those who took the test. This does not mean that the score was 65 percent out of 100, nor does it mean that 65 percent of the questions answered were correct. It means that the grade was higher than 65 percent of all those who took the test.

PERCENTILE: data divided into 100 equal parts

STANINE "standard nine" scores combine the understandability of percentages with the properties of the normal curve of probability. Stanines divide the bell curve into nine sections, the largest of which stretches from the 40th to the 60th percentile and is the "Fifth Stanine" (the average of taking into account error possibilities).

STANINE: scores that combine the understandability of percentages with the properties of the normal curve of probability

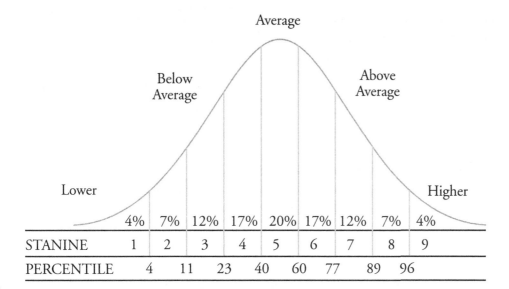

		4%	7%	12%	17%	20%	17%	12%	7%	4%	
STANINE		1	2	3	4	5	6	7	8	9	
PERCENTILE		4	11	23	40	60	77	89	96		

QUARTILES divide the data into four parts. First find the median of the data set (Q2), and then find the median of the upper (Q3) and lower (Q1) halves of the data set. There is some confusion in determining the upper and lower quartile and statisticians don't agree on which method to use. Tukey's method for finding the quartile values is to find the median of the data set, then find the median of the upper and lower halves of the data set. If there are an odd number of values in the data set, include the median value in both halves when finding the quartile values. For example, if we have the data set

> {1, 4, 9, 16, 25, 36, 49, 64, 81}

QUARTILES: data divided into four parts

First find the median value, which is 25. Since there are an odd number of values in the data set (9), we include the median in both halves. To find the quartile values, we must find the medians of:

> {1, 4, 9, 16, 25} and {25, 36, 49, 64, 81}

Since each of these subsets has an odd number of elements (5), we use the middle value. Thus, the lower quartile value is 9 and the upper quartile value is 49.

If the test you are taking allows the use of the TI-83, you should know that it uses a method described by Moore and McCabe (sometimes referred to as "M-and-M") to find quartile values. Their method is similar to Tukey's, but you don't include the median in either half when finding the quartile values. Using M-and-M on the data set above

> {1, 4, 9, 16, 25, 36, 49, 64, 81}

First find that the median value is 25. This time we'll exclude the median from each half. To find the quartile values, we must find the medians of

> {1, 4, 9, 16} and {36, 49, 64, 81}

Since each of these data sets has an even number of elements (4), we average the middle two values. Thus, the lower quartile value is $\frac{(4 + 9)}{2} = 6.5$, and the upper quartile value is $\frac{(49 + 64)}{2} = 56.5$.

With each of the preceding methods, the quartile values are always either one of the data points, or exactly half way between two data points.

Variance and Standard Deviation

An understanding of the definitions is important in determining the validity and uses of statistical data. All definitions and applications in this section apply to ungrouped data.

- Data item: each piece of data is represented by the letter X.

- Mean: the average of all data represented by the symbol $\overline{X}$.

- Range: difference between the highest and lowest value of data items.

- Sum of the squares: sum of the squares of the differences between each item and the mean. $Sx^2 = \text{Sum of } (X - \overline{X})^2$

- Variance: the sum of the squares quantity divided by the number of items. (The lowercase Greek letter sigma (σ) squared represents variance.) $\frac{Sx^2}{N} = \sigma^2$ The larger the value of the variance, the larger the spread.

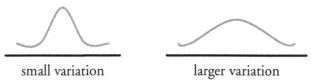

small variation   larger variation

- Standard deviation: the square root of the variance. The lowercase Greek letter sigma (σ) is used to represent standard deviation. $\sigma = \sqrt{\sigma^2}$

Most statistical calculators have standard deviation keys on them and should be used when asked to calculate statistical functions. It is important to become familiar with the calculator and the location of the keys needed.

Example: Given the ungrouped data below, calculate the mean, range, standard deviation, and variance.

15 22 28 25 34 38
18 25 30 33 19 23
Mean ($\overline{X}$) = 25.8333333
Range: 38 − 15 = 23
Standard deviation (σ) = 6.6991337
Variance (σ^2) = 48.87879

Finally, graphical representations of data sets, like box-and-whisker plots, help relate the measures of central tendency to data outliers, clusters and gaps. Consider the hypothetical box-and-whisker plot with one outlier value on each end of the distribution.

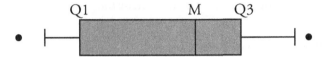

Note the beginning of the box is the value of the first quartile of the data set, and the end is the value of the third quartile. We represent the median as a vertical line in the box. The "whiskers" extend to the last point that is not an outlier (i.e., within $\frac{3}{2}$ times the range between Q1 and Q3 from either end of the box). The points beyond the figure represent outlier values.

COMPETENCY 4B
DISCRETE MATHEMATICS

Use and interpret statements that contain logical connectives (and, or, if–then), as well as logical quantifiers (some, all, none)

Logical Connectives and Conditional Statements

A SIMPLE STATEMENT represents a simple idea that can be described either as *true* or *false*, but not both. A small letter of the alphabet represents a simple statement.

Example: "Today is Monday."
This is a simple statement, since we can determine that this statement is either true or false. We can write $p =$ "Today is Monday."

Example: "Justin, please be quiet!"
We do not consider this a simple statement in our study of logic, since we cannot assign a truth value to it.

Simple statements joined by CONNECTIVES (*and, or, not, if–then,* and *if and only if*) result in compound statements. Note that we can also form compound statements using *but, however,* or *nevertheless.* We can assign a truth value to a compound statement.

SIMPLE STATEMENT: represents a simple idea that can be described either as true or false, but not both

CONNECTIVES: connectives (*and, or, not, if–then,* and *if and only if*)

HYPOTHESIS: known as the *if* clause of the conditional

CONCLUSION: called the *then* clause

We frequently write conditional statements in *if–then* form. The *if* clause of the conditional is known as the HYPOTHESIS, and the *then* clause is called the CONCLUSION. In a proof, the hypothesis is the information that is assumed true, while the conclusion is what is to be proven true. We consider a conditional to be of the form "if p, then q," where p is the hypothesis and q is the conclusion.

$p \rightarrow q$ is read, "if p, then q."

$\sim$ (statement) is read, "It is not true that (statement)."

Example: If an angle has a measure of 90 degrees, then it is a right angle.
In this statement, "an angle has a measure of 90 degrees" is the hypothesis, and "it is a right angle" is the conclusion.

Example: If you are in Pittsburgh, then you are in Pennsylvania.
In this statement, "you are in Pittsburgh" is the hypothesis, and "you are in Pennsylvania" is the conclusion.

QUANTIFIERS: words that describe a quantity under discussion

Logical Quantifiers and Negations

QUANTIFIERS are words that describe a quantity under discussion. These include words such as *all, none* (or *no*), and *some.*

NEGATION of a statement: If a statement is true, then its negation must be false (and vice versa).

NEGATION: if a statement is true, then its negation must be false (and vice versa)

NEGATION RULES	
Statement	**Negation**
q	not q
not q	q
π and s	(not π) or (not s)
π or s	(not π) and (not s)
if p, then q	(p) and (not q)

Example: Select the statement that is the negation of "Some winter nights are not cold."

A. All winter nights are not cold.

B. Some winter nights are cold.

C. All winter nights are cold.

D. None of the winter nights is cold.

The negation of some are is none is. Therefore, the negation statement is "None of the winter nights is not cold." This is equivalent to statement C. Therefore the answer is C.

Example: Select the statement that is the negation of "If it rains, then the beach party will be called off."

A. If it does not rain, then the beach party will not be called off.

B. If the beach party is called off, then it will not rain.

C. It does not rain, and the beach party will not be called off.

D. It rains, and the beach party will not be called off.

The negation of "If p, then q" is "p and (not q)." The negation of the given statement is "It rains, and the beach party will not be called off." Select D.

Example: Select the negation of the statement "If they are elected, then all politicians go back on election promises."

A. If they are elected, then many politicians go back on election promises.

B. They are elected and some politicians go back on election promises.

C. If they are not elected, some politicians do not go back on election promises.

D. None of the above statements is the negation of the given statement.

Identify the key words of "if ... then" and "all ... go back." The negation of the given statement is "They are elected and none of the politicians goes back on election promises." So select response D, since statements A, B, and C are not the negations.

Example: Select the statement that is the negation of "The sun is shining bright, and I feel great."

A. If the sun is not shining bright, I do not feel great.

B. The sun is not shining bright, and I do not feel great.

C. The sun is not shining bright, or I do not feel great.

D. The sun is shining bright, and I do not feel great.

The negation of "r and s" is "(not r) or (not s)." Therefore, the negation of the given statement is "The sun is *not* shining bright, *or* I do not feel great." We select response C.

Inverse, Converse, and Contrapositive

LOGICAL COMPARISONS	
Conditional: If *p*, then *q*	*p* is the hypothesis, and *q* is the conclusion.
Inverse: If ~ *p*, then ~ *q*	Negate both the hypothesis and the conclusion from the original conditional. (If not *p*, then not *q*.)
Converse: If *q*, then *p*	Reverse the two clauses. The original hypothesis becomes the conclusion. The original conclusion then becomes the new hypothesis.
Contrapositive: If ~ *q*, then ~ *p*	Reverse the two clauses. The original hypothesis becomes the conclusion. The original conclusion then becomes the new hypothesis. Then negate both the new hypothesis and the new conclusion.

Example: Given the **conditional***:*

> If an angle has 60°, then it is an acute angle.

Its **inverse**, in the form "If ~ *p*, then ~ *q*," would be:

> If an angle doesn't have 60°, then it is not an acute angle.

Notice that the inverse is false, even though the conditional statement was true.

Its **converse**, in the form "If *q*, then *p*," would be:

> If an angle is an acute angle, then it has 60°.

Notice that the converse is false, even though the conditional statement was true.

Tip: *If you are asked to pick a statement that is logically equivalent to a given conditional, look for the contrapositive. The inverse and converse are not always logically equivalent to every conditional. The contrapositive is always logically equivalent.*

Its **contrapositive**, in the form "If *q*, then *p*," would be:

> If an angle isn't an acute angle, then it doesn't have 60°.

Notice that the contrapositive is true, assuming the original conditional statement was true.

Example: Find the inverse, converse, and contrapositive of the following conditional statements. Also determine whether each of the four statements is true or false.

Conditional: If $x = 5$ then $x^2 - 25 = 0$.	True
Inverse: If $x \neq 5$, then $x^2 - 25 \neq 0$.	False, x could be –5
Converse: If $x^2 - 25 = 0$, then $x = 5$	False, x could be –5
Contrapositive: If $x^2 - 25 \neq 0$, then $x \neq 5$.	True
Conditional: If $x = 5$, then $6x = 30$.	True
Inverse: If $x \neq 5$, then $6x \neq 30$.	True
Converse: If $6x = 30$, then $x = 5$.	True
Contrapositive: If $6x \neq 30$, then $x \neq 5$	True

Sometimes, as in this example, all four statements can be logically equivalent, but the only statement that will always be logically equivalent to the original conditional is the contrapositive.

Venn Diagrams

We can diagram conditional statements using a VENN DIAGRAM. We can draw a diagram with one figure inside another figure. The inner figure represents the hypothesis, and the outer figure represents the conclusion. If we take the hypothesis to be true, then you are located inside the inner figure. If you are located in the inner figure, then you are also inside the outer figure, so that proves that the conclusion is true.

> **VENN DIAGRAM:** a diagram using conditional statements

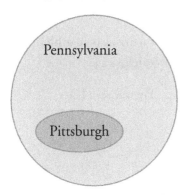

Sometimes that conclusion can then be used as the hypothesis for another conditional, which can result in a second conclusion.

Suppose the following statements were given to you, and you were asked to try to reach a conclusion:

A. All swimmers are athletes.
  All athletes are scholars.

  In "if-then" form, these would be:
  If you are a swimmer, then you are an athlete.
  If you are an athlete, then you are a scholar.

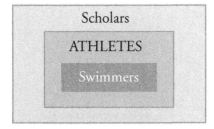

Clearly, if you are a swimmer, then you are also an athlete. This includes you in the group of scholars.

B. All swimmers are athletes.
All wrestlers are athletes.

In "if-then" form, these would be:
If you are a swimmer, then you are an athlete.
If you are a wrestler, then you are an athlete.

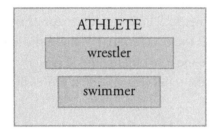

Clearly, if you are a swimmer or a wrestler, then you are also an athlete. This does NOT allow you to come to any other conclusions.

A swimmer may or may NOT also be a wrestler. Therefore, NO CONCLUSION IS POSSIBLE.

C. All rectangles are parallelograms.
Quadrilateral ABCD is not a parallelogram.

In "if–then" form, the first statement would be:
If a figure is a rectangle, then it is also a parallelogram.

Note that the second statement is the negation of the conclusion of statement one. Remember also that the contrapositive is logically equivalent to a given conditional—that is, "*If ~ q, then ~ p.*" Since "ABCD is not a parallelogram" is like saying "*If ~ q,*" then you can come to the conclusion "*then ~ p.*" Therefore, the conclusion is ABCD is not a rectangle.

Looking at the following Venn diagram, if all rectangles are parallelograms, then rectangles are included as part of the parallelograms. Since quadrilateral ABCD is not a parallelogram, that it is excluded from anywhere inside the parallelogram box. This allows you to conclude that ABCD cannot be a rectangle either.

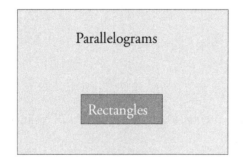

Counterexamples

A COUNTEREXAMPLE is an exception to a proposed rule or conjecture that disproves the conjecture. For example, the existence of a single non-brown dog disproves the conjecture "All dogs are brown." Thus, any non-brown dog is a counterexample.

In searching for mathematical counterexamples, one should consider extreme cases near the ends of the domain of an experiment and special cases where an additional property is introduced. Examples of extreme cases are numbers near zero and obtuse triangles that are nearly flat. An example of a special case for a problem involving rectangles is a square because a square is a rectangle with the additional property of symmetry.

Example: Identify a counterexample for the following conjectures.

1. If n is an even number, then $n + 1$ is divisible by 3.

   $n = 4$

   $n + 1 = 4 + 1 = 5$

   5 is not divisible by 3.

2. If n is divisible by 3, then $n^2 - 1$ is divisible by 4.

   $n = 6$

   $n^2 - 1 = 6^2 - 1 = 35$

   35 is not divisible by 4.

> **COUNTEREXAMPLE:** an exception to a proposed rule or conjecture that disproves the conjecture

---

SKILL 4B.2 **Solve problems involving the union and intersection of sets, subsets, and disjoint sets**

Properties of Sets

Set theory is a helpful tool for organizing and describing information in mathematics. Any collection of elements can be considered a set. For instance, a set could be as simple as {banana, apple, pear} or as complicated as $\mathbb{C}$, the set of complex numbers. Two basic set operations are union (symbolized by $\cup$) and intersection (symbolized by $\cap$). Two sets that have no members in common are disjoint.

The UNION ($\cup$) of two sets is the set of all the numbers that are in either the first or the second set or in both sets.

   If set A = {-7, -2, 0, 1, 3, 4, 5, 6} and set B = {-5, -3, 0, 1, 2, 3, 5}, then A $\cup$ B is {-7, -5, -3, -2, 0, 1, 2, 3, 4, 5, 6}.

> **UNION:** in two sets, the set of all the numbers, which are in either the first or the second set or in both sets

The INTERSECTION (∩) of two sets is the set of numbers that are in both sets.

A(∩) B is{0, 1, 3, 5}.

INTERSECTION: in two sets of numbers, the numbers that are in both sets

We can illustrate the union and intersection of sets A and B with a Venn diagram.

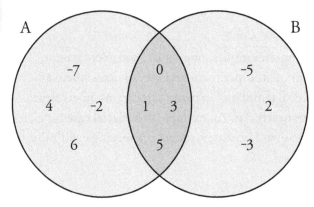

The union of the two sets would be the entire Venn diagram, whereas the intersection would be the area in the middle of the diagram where the two circles overlap.

NULL SET: called the empty set

The NULL SET is also called the empty set; it is the set that does not contain any numbers. The null set can be expressed two different ways: either { } or ∅. {0} is not the null set, since it does have one element.

DISJOINT SETS: sets that have no elements in common

If sets A and B did not contain 0, 1, 3, and 5, they would be disjoint sets. DISJOINT SETS are sets that have no elements in common. They would look like this:

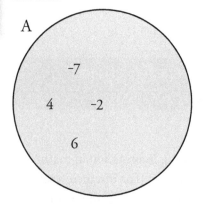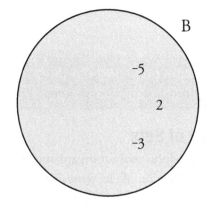

Example: Find the union of {a, b, c} and {d, e, f}.
Solution: {a, b, c} ∪{d, e, f} = {a, b, c, d, e, f}

Example: Find the intersection of {a, b, c, d} and {b, d, e, f}.
Solution: {a, b, c, d} ∩ {b, d, e, f} = {b, d}

SUBSET: set A is a subset of set B if every element of set A is an element of set B

Set A is a SUBSET of set B if every element of set A is an element of set B.

Example: Set A = {Virginia, Rhode Island}; Set B = {the United States}
A ⊆ B

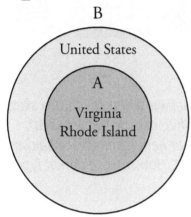

The result of a union of two sets contains all the elements found in each set. Thus, for instance, the union of the set of irrational numbers and the set of rational numbers is the set of real numbers.

The result of an intersection of two sets contains all the elements that are common to both sets. The intersection of the sets of irrational and rational numbers is an empty set (or null set) because there are no numbers that are both rational and irrational (the two sets are disjoint).

Example: Find the union and intersection of the sets of real and complex numbers.

The set of real numbers is expressed as $\mathbb{R}$, and the set of complex numbers is expressed as $\mathbb{C}$. Since the set of real numbers is contained in the set of complex numbers, the following can be written:

$\mathbb{R} \cup \mathbb{C} = \mathbb{C}$

For the intersection, only the real numbers are common to both sets. Thus:

$\mathbb{R} \cap \mathbb{C} = \mathbb{R}$

Sets may have a limited number of elements or an infinite number.

The ordering of the elements in a set is not relevant, although a certain order may be imposed for convenience when writing all or some portion of the elements of a set. For instance, the set of the first four natural numbers can be equivalently written {1, 2, 3, 4} or {4, 2, 1, 3} (or any other arrangement).

Sets may also be written in a conditional form that specifies a certain collection of elements. For instance, $\{x : x \geq 3\}$ specifies the set of x such that x is greater than or equal to 3. Another example is $\{x : x \in \mathbb{Z}\}$, which is the set of integers (this set could be written more simply as just $\mathbb{Z}$).

> **SKILL 4B.3** Solve basic counting problems involving permutations and combinations without necessarily knowing formulas; use Pascal's triangle to solve problems

Counting techniques are discussed in Skill 4A.3

Example: Suppose you want to order a pizza. You have a choice of three sizes (small, medium, or large), three types of crust (thin, pan, or hand-tossed), four choices of meat (pepperoni, sausage, both, or none), and three choices of cheese (regular, double, or stuffed with cheese). How many different types of pizza could you order?

$3 \times 3 \times 4 \times 3 = 108$ different types of pizza

Another method of basic counting is the tree diagram.

Example: Suppose you want to look at the possible sequence of events for having two children in a family. Since a child will be either a boy or a girl, you would have the following tree diagram to illustrate the possible outcomes:

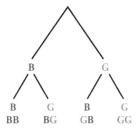

From the diagram, you see that there are four possible outcomes, two of which are the same.

Using Pascal's Triangle to Solve Problems

Pascal's triangle looks like the following:

```
            1
          1   1
        1   2   1
      1   3   3   1
    1   4   6   4   1
  1   5  10  10   5   1
```

where the sum of Row 0 = 2^0, Row 1 = 2^1, Row 2 = 2^2, Row 3 = 2^3, Row 4 = 2^4, Row 5 = 2^5. From this, we can see that the sum of row n would be 2^n.

Pascal's Triangle is useful in experiments in which there are only two equally likely possibilities, such as coin tosses.

Example: Find the probability of getting at least 5 heads when tossing 6 coins.

Solution: The fundamental counting property tells us that there are 2^6 possible outcomes when tossing 6 coins. We then construct the row of Pascal's Triangle that begins 1, 6:

1   6   15   20   15   6   1

There are 6 ways of getting 5 heads and 1 way of getting 6 heads. Therefore, we find the probability of tossing at least 5 heads with 6 coins as:

$$\frac{(1+6)}{2^6} = \frac{7}{64}$$

> **SKILL 4B.4** Solve problems that involve simple sequences or number patterns *(e.g., triangular numbers or other geometric numbers)*; find rules for number patterns

Problems that involve sequences or number patterns

The most common NUMERICAL PATTERNS are arithmetic sequences and geometric sequences. In an arithmetic sequence, each term is separated from the next by a fixed number (e.g., 3, 6, 9, 12, 15, . . .). In a geometric sequence, each term in the series is multiplied by a fixed number to get the next term (e.g., 3, 6, 12, 24, 48, . . .).

> **NUMERICAL PATTERNS:** arithmetic sequences and geometric sequences

Arithmetic sequences

An ARITHMETIC SEQUENCE is a set of numbers with a common difference between the terms. Terms and the distance between terms can be calculated using the following formula:

$a_n = a_1 + (n-1)d$, where
a_1 = the first term
a_n = the nth term (general term)
n = the number of the term in the sequence
d = the common difference

> **ARITHMETIC SEQUENCE:** a set of numbers with a common difference between the terms

The formula essentially expresses the arithmetic sequence as an algebraic pattern $a_1, a_1 + d, a_1 + 2d, a_1 + 3d$, and so on, where any numbers can be substituted for a1 and d to derive different numerical sequences.

Example: Find the 8th term of the arithmetic sequence 5, 8, 11, 14, . . .

$a_n = a_1 + (n - 1)d$

$a_1 = 5$ Identify the 1st term.

$d = 8 - 5 = 3$ Find d.

$a_8 = 5 + (8{-}1)3$ Substitute.

$a_8 = 26$

Example: Given two terms of an arithmetic sequence, find a_1 and d.

$a_4 = 21$	$a_6 = 32$
$a_n = a_1 + (n - 1)d$	$a_4 = 21, n = 4$
$21 = a_1 + (4 - 1)d$	$a_6 = 32, n = 6$
$32 = a_1 + (6 - 1)d$	
$21 = a_1 + 3d$	Solve the system of equations.
$32 = a_1 + 5d$	

$32 = a_1 + 5d$

$\underline{-21 = \text{-}a_1 - 3d}$ Multiply by –1.

$11 = \qquad 2d$ Add the equations.

$5.5 = d$

$21 = a_1 + 3(5.5)$ Substitute d = 5.5 into either equation.

$21 = a_1 + 16.5$

$a_1 = 4.5$

The sequence begins with 4.5 and has a common difference of 5.5 between numbers.

Geometric sequences

A GEOMETRIC SEQUENCE is a series of numbers in which a common ratio can be multiplied by a term to yield the next term. The common ratio can be calculated using the formula:

$r = \dfrac{a_{n+1}}{a_n}$, where

r = common ratio

a_n = the nth term

> **GEOMETRIC SEQUENCE:** a series of numbers in which a common ratio can be multiplied by a term to yield the next term

The ratio is then used in the geometric sequence formula:

$a_n = a_1 r^{n-1}$

The formula essentially expresses the geometric sequence as an algebraic pattern $a_1, a_1 r, a_1 r^2, a_1 r^3, a_1 r^4$, and so on, where any numbers can be substituted for a_1 and r to derive different numerical sequences.

Example: Find the 8th term of the geometric sequence 2, 8, 32, 128, . . .

$r = \frac{a_{n+1}}{a_n}$ Use the common ratio formula to find the ratio.

$r = \frac{8}{2}$ Substitute $a_n = 2$, $a_{n+1} = 8$.

$r = 4$

$a_n = a_1 \times r^{n-1}$ Use $r = 4$ to solve for the 8th term.

$a_n = 2 \times 4^{8-1}$

$a_n = 32{,}768$

Recursive Patterns

A RECURRENCE RELATION is an equation that defines a sequence recursively; in other words, each term of the sequence is defined as a function of the preceding terms. For instance, the formula for the balance of a mortgage principal after i payments, can be expressed recursively as follows.

$A_i = A_{i-1}(1 + \frac{r}{n}) - M$ where $A_0 = P$ which is the initial principal invested.

Although this formula is not helpful as it stands for directly calculating parameters of a mortgage, it can be helpful for calculating balances (for example) given a specific set of circumstances somewhere within a mortgage term. Compound interest and annuities can also be expressed in recursive form.

Calculation of a past or future term by applying the recursive formula multiple times is called ITERATION.

Sequences of numbers can be defined by iteratively applying a recursive pattern. For instance, the Fibonacci sequence is defined as follows.

$F_i = F_{i-1} + F_{i-2}$, where $F_0 = 0$ and $F_1 = 1$

Applying this recursive formula gives the sequence $\{0, 1, 1, 2, 3, 5, 8, 13, 21, \ldots\}$.

It is sometimes difficult or impossible to write recursive relations in explicit or closed form. In such cases, especially where computer programming is involved, the recursive form can still be helpful. When the elements of a sequence of numbers or values depend on one or more previous values, then it is possible that a recursive formula could be used to summarize the sequence.

If a value or number from a later point in the sequence (that is, other than the beginning) is known and it is necessary to find previous terms, then the indices of the recursive relation can be adjusted to find previous values instead of later ones. Consider, for instance, the Fibonacci sequence.

$F_i = F_{i-1} + F_{i-2}$

$F_{i+2} = F_{i+1} + F_i$

$F_i = F_{i+2} - F_{i+1}$

RECURRENCE RELATION: an equation that defines a sequence recursively

ITERATION: calculation of a past or future term by applying the recursive formula multiple times

Thus, if any two consecutive numbers in the Fibonacci sequence are known, then the previous numbers of the sequence can be found (in addition to the later numbers).

Example: Write a recursive formula for the following sequence: {2, 3, 5, 9, 17, 33, 65, …}.

By inspection, it can be seen that each number in the sequence is equal to twice the previous number, less one. If the numbers in the sequence are indexed such that for the first number $i = 1$ and so on, then the recursion relation is the following.

$$N_i = 2N_{i-1} - 1$$

Example: If a recursive relation is defined by $N_i = N_{i-1}^2$ and the fourth term is 65,536, what is the first term?

Adjust the indices of the recursion and then solve for N_i.

$$N_{i+1} = N_i^2$$
$$N_i = \sqrt{N_{i+1}}$$

Use this relationship to backtrack to the first term.

$$N_3 = \sqrt{N_4} = \sqrt{65,536} = 256$$
$$N_2 = \sqrt{N_3} = \sqrt{256} = 16$$
$$N_1 = \sqrt{N_2} = \sqrt{16} = 4$$

The first term of the sequence is thus 4.

**SKILL Use and interpret matrices as tools for displaying data
4B.5**

Properties of Matrices

MATRIX: an array of an ordered set of numbers called elements

DIMENSIONS: of a matrix are written as the number of rows by the number of columns ($r \times c$)

A MATRIX is an array of an ordered set of numbers called elements. An example matrix is shown below. The DIMENSIONS of a matrix are written as the number of rows by the number of columns ($r \times c$).

$$\begin{pmatrix} 0 & 3 & 1 \\ 4 & 2 & 3 \\ 1 & 0 & 2 \end{pmatrix}$$

Since this matrix has three rows and three columns, it is called a 3×3 matrix. The element in the second row of the third column would be denoted as $3_{2,3}$.

$$\begin{pmatrix} 1 & 2 & 3 \\ 4 & 5 & 6 \end{pmatrix} \quad \text{is a } 2 \times 3 \text{ matrix (2 rows by 3 columns)}$$

$$\begin{pmatrix} 1 & 2 \\ 3 & 4 \\ 5 & 6 \end{pmatrix}$$ is a 3 × 2 matrix (3 rows by 2 columns)

Matrices can be added or subtracted only if their dimensions are the same. To add (subtract) compatible matrices, simply add (subtract) the corresponding elements, as with the example below for 2 × 2 matrices.

$$\begin{pmatrix} a_{11} & a_{12} \\ a_{21} & a_{22} \end{pmatrix} + \begin{pmatrix} b_{11} & b_{12} \\ b_{21} & b_{22} \end{pmatrix} = \begin{pmatrix} a_{11} + b_{11} & a_{11} + b_{12} \\ a_{21} + b_{21} & a_{22} + b_{22} \end{pmatrix}$$

Determinants

Associated with every square matrix is a number called the determinant.

Use these formulas to calculate determinants.

$$2 \times 2 \begin{pmatrix} a & b \\ c & d \end{pmatrix} = ad - bc$$

$$3 \times 3$$

$$\begin{pmatrix} a_1 & b_1 & c_1 \\ a_2 & b_2 & c_2 \\ a_3 & b_3 & c_3 \end{pmatrix} = (a_1 b_2 c_3 + b_1 c_2 a_3 + c_1 a_2 b_3) - (a_3 b_2 c_1 + b_3 c_2 a_1 + c_3 a_2 b_1)$$

This formula is found by repeating the first two columns and then using the diagonal lines to find the value of each expression as shown below:

$$\begin{pmatrix} a_1^* & b_1^\circ & c_1^\bullet \\ a_2 & b_2^* & c_2^\circ \\ a_3 & b_3 & c_3^* \end{pmatrix} \begin{matrix} a_1 & b_1 \\ a_2^\bullet & b_2 \\ a_3^\circ & b_3^\bullet \end{matrix}$$
$$= (a_1 b_2 c_3 + b_1 c_2 a_3 + c_1 a_2 b_3) - (a_3 b_2 c_1 + b_3 c_2 a_1 + c_3 a_2 b_1)$$

Example: Find the value of the determinant:

$$\begin{pmatrix} 4 & -8 \\ 7 & 3 \end{pmatrix} = (4)(3) - (7)(-8) \quad \text{Cross-multiply and subtract.}$$

$$12 - (-56) = 68 \qquad \text{Then simplify.}$$

Adding and Subtracting Matrices

Addition of matrices is accomplished by adding the corresponding elements of the two matrices. Subtraction is defined as the inverse of addition. In other words, change the sign on all the elements in the second matrix and add the two matrices.

Example: Find the sum.

$$\begin{pmatrix} 2 & 3 \\ -4 & 7 \\ 8 & -1 \end{pmatrix} + \begin{pmatrix} 8 & -1 \\ 2 & -1 \\ 3 & -2 \end{pmatrix} =$$

$$\begin{pmatrix} 2+8 & 3+(-1) \\ -4+2 & 7+(-1) \\ 8+3 & -1+(-2) \end{pmatrix} \qquad \text{Add corresponding elements.}$$

$$\begin{pmatrix} 10 & 2 \\ -2 & 6 \\ 11 & -3 \end{pmatrix} \qquad \text{Simplify.}$$

Example: Find the difference.

$$\begin{pmatrix} 8 & -1 \\ 7 & 4 \end{pmatrix} - \begin{pmatrix} 3 & 6 \\ -5 & 1 \end{pmatrix} =$$

$$\begin{pmatrix} 8 & -1 \\ 7 & 4 \end{pmatrix} + \begin{pmatrix} -3 & -6 \\ 5 & -1 \end{pmatrix} =$$

Change all of the signs in the second matrix and then add the two matrices.

$$\begin{pmatrix} 8+(-3) & -1+(-6) \\ 7+5 & 4+(-1) \end{pmatrix} = \qquad \text{Simplify.}$$

$$\begin{pmatrix} 5 & -7 \\ 12 & 3 \end{pmatrix}$$

Practice Set Problems

1. $\begin{pmatrix} 8 & -1 \\ 5 & 3 \end{pmatrix} + \begin{pmatrix} 3 & 8 \\ 6 & -2 \end{pmatrix} =$

2. $\begin{pmatrix} 3 & 7 \\ -4 & 12 \\ 0 & -5 \end{pmatrix} - \begin{pmatrix} 3 & 4 \\ 6 & -1 \\ -5 & -5 \end{pmatrix} =$

Answer Key

1. $\begin{pmatrix} 11 & 7 \\ 11 & 1 \end{pmatrix}$

2. $\begin{pmatrix} 0 & 3 \\ -10 & 13 \\ 5 & 0 \end{pmatrix}$

SCALAR MULTIPLICA-TION: the product of the scalar (the outside number) and each element inside the matrix

Scalar Multiplication of Matrices

SCALAR MULTIPLICATION is the product of the scalar (the outside number) and each element inside the matrix.

Example: Given:

$A = \begin{pmatrix} 4 & 0 \\ 3 & -1 \end{pmatrix}$, find 2A.

$$2A = 2 \begin{pmatrix} 4 & 0 \\ 3 & -1 \end{pmatrix}$$

$\begin{pmatrix} 2 \times 4 & 2 \times 0 \\ 2 \times 3 & 2 \times -1 \end{pmatrix}$ Multiply each element in the matrix by the scalar.

$\begin{pmatrix} 8 & 0 \\ 6 & -2 \end{pmatrix}$ Simplify.

Practice Set Problems

1. $-2 \begin{pmatrix} 2 & 0 & 1 \\ -1 & -2 & 4 \end{pmatrix} =$

2. $3 \begin{pmatrix} 6 \\ 2 \\ 8 \end{pmatrix} + 4 \begin{pmatrix} 0 \\ 7 \\ 2 \end{pmatrix}$

3. $2 \begin{pmatrix} -6 & 8 \\ -2 & -1 \\ 0 & 3 \end{pmatrix}$

Answer Key

1. $\begin{pmatrix} -4 & 0 & -2 \\ 2 & 4 & -8 \end{pmatrix}$

2. $\begin{pmatrix} 18 \\ 34 \\ 32 \end{pmatrix}$

3. $\begin{pmatrix} -12 & 16 \\ -4 & -2 \\ 0 & 6 \end{pmatrix}$

The variable in a matrix equation represents a matrix. When solving for the answer use the adding, subtracting, and scalar multiplication properties.

Example: Solve the matrix equation for the variable x.

$2x + \begin{pmatrix} 4 & 8 & 2 \\ 7 & 3 & 4 \end{pmatrix} = 2 \begin{pmatrix} 1 & -2 & 0 \\ 3 & -5 & 7 \end{pmatrix}$

$2x = 2 \begin{pmatrix} 1 & -2 & 0 \\ 3 & -5 & 7 \end{pmatrix} - \begin{pmatrix} 4 & 8 & 2 \\ 7 & 3 & 4 \end{pmatrix}$ Subtract $\begin{pmatrix} 4 & 8 & 2 \\ 7 & 3 & 4 \end{pmatrix}$ from both sides.

$$2x = \begin{pmatrix} 2 & -4 & 0 \\ 6 & -10 & 14 \end{pmatrix} + \begin{pmatrix} -4 & -8 & -2 \\ -7 & -3 & -4 \end{pmatrix}$$ Scalar multiplication and matrix subtraction.

$$2x = \begin{pmatrix} -2 & -12 & -2 \\ -1 & -13 & 10 \end{pmatrix}$$ Matrix addition.

$$x = \begin{pmatrix} -1 & -6 & -1 \\ -\frac{1}{2} & -\frac{13}{2} & 5 \end{pmatrix}$$ Multiply both sides by $\frac{1}{2}$ or divide by 2.

Example: Solve for the unknown values of the elements in the matrix.

$$\begin{pmatrix} x+3 & y-2 \\ z+3 & w-4 \end{pmatrix} + \begin{pmatrix} -2 & 4 \\ 2 & 5 \end{pmatrix} = \begin{pmatrix} 4 & 8 \\ 6 & 1 \end{pmatrix}$$

$$\begin{pmatrix} x+1 & y+2 \\ z+5 & w+1 \end{pmatrix} = \begin{pmatrix} 4 & 8 \\ 6 & 1 \end{pmatrix}$$ Matrix addition.

$x + 1 = 4 \quad y + 2 = 8 \quad z + 5 = 6 \quad w + 1 = 1$

$x = 3 \qquad\quad y = 6 \qquad\quad z = 1 \qquad\quad w = 0$

Definition of equal matrices.

Practice Set Problems

1.
$$x + \begin{pmatrix} 7 & 8 \\ 3 & -1 \\ 2 & -3 \end{pmatrix} = \begin{pmatrix} 0 & 8 \\ -9 & -4 \\ 8 & 2 \end{pmatrix}$$

2.
$$4x - 2\begin{pmatrix} 0 & 10 \\ 6 & -4 \end{pmatrix} = 3\begin{pmatrix} 4 & 9 \\ 0 & 12 \end{pmatrix}$$

3.
$$\begin{pmatrix} 7 & 3 \\ 2 & 4 \\ 3 & 7 \end{pmatrix} + \begin{pmatrix} a+2 & b+4 \\ c-3 & d+1 \\ e & f+3 \end{pmatrix} = \begin{pmatrix} 4 & 6 \\ -1 & 1 \\ 3 & 0 \end{pmatrix}$$

Answer Key

1.
$$\begin{pmatrix} -7 & 0 \\ -12 & -3 \\ 6 & 5 \end{pmatrix}$$

2.
$$\begin{pmatrix} 3 & \frac{47}{4} \\ 3 & 7 \end{pmatrix}$$

3. $a = -5 \quad b = -1 \quad c = 0 \quad d = -4$
$e = 0 \quad f = -10$

The product of two matrices can only be found if the number of columns in the first matrix is equal to the number of rows in the second matrix.

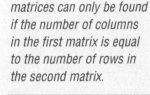

Matrix Multiplication

The product of two matrices can only be found if the number of columns in the first matrix is equal to the number of rows in the second matrix. Matrix multiplication is not necessarily commutative.

Examples:

Find the product *AB* if:

$$A = \begin{pmatrix} 2 & 3 & 0 \\ 1 & -4 & -2 \\ 0 & 1 & 1 \end{pmatrix} \qquad B = \begin{pmatrix} -2 & 3 \\ 6 & -1 \\ 0 & 2 \end{pmatrix}$$

$$3 \times 3 \qquad\qquad 3 \times 2$$

Note: Since the number of columns in the first matrix ($3 \times \underline{3}$) matches the number of rows in the second matrix ($\underline{3} \times 2$), this product is defined and can be found. The dimensions of the product will be equal to the number of rows in the first matrix ($\underline{3} \times 3$) by the number of columns in the second matrix ($3 \times \underline{2}$). The answer will be a 3×2 matrix.

$$AB = \begin{pmatrix} 2 & 3 & 0 \\ 1 & -4 & -2 \\ 0 & 1 & 1 \end{pmatrix} \times \begin{pmatrix} -2 & 3 \\ 6 & -1 \\ 0 & 2 \end{pmatrix}$$

$$\begin{pmatrix} 2(-2) + 3(6) + 0(0) \\ \\ \end{pmatrix}$$ Multiply the first row of *A* by the first column of *B*.

$$\begin{pmatrix} 14 & 2(3) + 3(-1) + 0(2) \\ \\ \end{pmatrix}$$ Multiply the first row of *A* by the second column of *B*.

$$\begin{pmatrix} 14 & 3 \\ 1(-2) - 4(6) - 2(0) & \\ \end{pmatrix}$$ Multiply the second row of *A* by the first column of *B*.

$$\begin{pmatrix} 14 & 3 \\ -26 & 1(3) - 4(-1) - 2(2) \\ \end{pmatrix}$$ Multiply the second row of *A* by the second column of *B*.

$$\begin{pmatrix} 14 & 3 \\ -26 & 3 \\ 0(-2) + 1(6) + 1(0) & \end{pmatrix}$$ Multiply the third row of *A* by the first column of *B*.

$$\begin{pmatrix} 14 & 3 \\ -26 & 3 \\ 6 & 0(3) + 1(-1) + 1(2) \end{pmatrix}$$ Multiply the third row of *A* by the second column of *B*.

$$\begin{pmatrix} 14 & 3 \\ -26 & 3 \\ 6 & 1 \end{pmatrix}$$

The product of *BA* is not defined, since the number of columns in *B* is not equal to the number of rows in *A*.

Practice Set Problems

1. $\begin{pmatrix} 3 & 4 \\ -2 & 1 \end{pmatrix}\begin{pmatrix} -1 & 7 \\ -3 & 1 \end{pmatrix}$

2. $\begin{pmatrix} 1 & -2 \\ 3 & 4 \\ 2 & 5 \\ -1 & 6 \end{pmatrix}\begin{pmatrix} 3 & -1 & -4 \\ -1 & 2 & 3 \end{pmatrix}$

Answer Key

1. $\begin{pmatrix} -15 & 25 \\ -1 & -13 \end{pmatrix}$

2. $\begin{pmatrix} 5 & -5 & -10 \\ 5 & 5 & 0 \\ 1 & 8 & 7 \\ -9 & 13 & 22 \end{pmatrix}$

Using Matrices to Solve Systems of Equations

When given the following system of equations:

$ax + by = e$
$cx + dy = f$

the matrix equation is written in the form:

$$\begin{pmatrix} a & b \\ c & d \end{pmatrix}\begin{pmatrix} x \\ y \end{pmatrix} = \begin{pmatrix} e \\ f \end{pmatrix}$$

The solution is found using the inverse of the matrix of coefficients. The inverse of matrices can be written as follows:

$$A^{-1} = \frac{1}{\text{determinant of } A}\begin{pmatrix} d & -b \\ -c & a \end{pmatrix}$$

Example: Write the matrix equation of the system, and solve the system.

$3x - 4y = 2$
$2x + y = 5$

$\begin{pmatrix} 3 & -4 \\ 2 & 1 \end{pmatrix}\begin{pmatrix} x \\ y \end{pmatrix} = \begin{pmatrix} 2 \\ 5 \end{pmatrix}$ Definition of matrix equation.

$\begin{pmatrix} x \\ y \end{pmatrix} = \frac{1}{11}\begin{pmatrix} 1 & 4 \\ -2 & 3 \end{pmatrix}\begin{pmatrix} 2 \\ 5 \end{pmatrix}$ Multiply by the inverse of the coefficient matrix.

$\begin{pmatrix} x \\ y \end{pmatrix} = \frac{1}{11}\begin{pmatrix} 22 \\ 11 \end{pmatrix}$ Matrix multiplication.

$$\begin{pmatrix} x \\ y \end{pmatrix} = \begin{pmatrix} 2 \\ 1 \end{pmatrix}$$ Scalar multiplication.

The solution is (2, 1).

Practice Set Problems

Solve each system.
1. $x + 2y = 5$
   $3x + 5y = 14$

Answer Key

1. $\begin{pmatrix} x \\ y \end{pmatrix} = \begin{pmatrix} 3 \\ 1 \end{pmatrix}$

The following is a simple example of how matrices may be used to display, interpret and perform operations on data.

Example: A company has two stores. The income and expenses (in dollars) for the two stores, for three months, are shown in the matrices.

April	Income	Expenses
Store 1	190,000	170,000
Store 2	100,000	110,000

May	Income	Expenses
Store 1	210,000	200,000
Store 2	125,000	120,000

June	Income	Expenses
Store 1	220,000	215,000
Store 2	130,000	115,000

The owner wants to know what his first-quarter income and expenses were, so he adds the three matrices.

1st Quarter	Income	Expenses
Store 1	620,000	585,000
Store 2	355,000	345,000

Then, to find the profit for each store:

Profit for Store 1 = $620,000 − $585,000 = $35,000
Profit for Store 2 = $355,000 − $345,000 = $10,000

Draw conclusions from information contained in simple diagrams, flowcharts, paths, circuits, networks, or algorithms

Diagrams

At all levels of mathematics, people use diagrams as an aid to visualizing and solving problems. There are many examples throughout this guide. Two more examples are given below.

Geometric problems lend themselves very easily to diagrammatic representation:

Example: An open swimming pool 50 ft long and 20 ft wide has a wide walkway of uniform breadth running all around it. If the area of the walkway is four times that of the pool, what is the outer perimeter of the walkway?

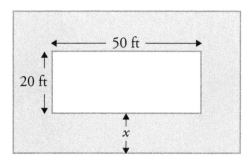

The diagram makes it easy to write the equations to represent the problem:

Let the width of the walkway be x ft.

Then, the length of the large rectangle $= 50 + 2x$ ft

The width of the large rectangle $= 20 + 2x$ ft

The area of the swimming pool $= 20 \times 50 = 1000$ sq. ft

The area of the large rectangle including the walkway and swimming pool $= (20 + 2x)(50 + 2x)$ sq. ft

Since the area of the walkway is 4 times that of the swimming pool, the area of the large rectangle including both the walkway and swimming pool is 5 times that of the swimming pool.

This relationship maybe expressed using the following quadratic equation:

$(20 + 2x)(50 + 2x) = 5000$

Even problems without geometric elements can be represented using diagrams.

Example: Mary is 5 years older than Jennifer who is 8 years younger than Sarah.

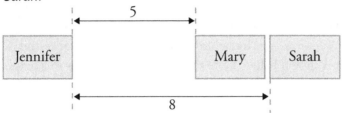

Flowcharts

A flowchart is a kind of diagram that represents the sequential steps in an algorithm or a process. Flowcharts are commonly used in computer programming. In general, a flow chart is typically used to represent a process that includes decision points where the process branches out into alternate paths.

Example: Create a flow chart to represent plans for a field day at a school. Morning classes will be followed by a pizza lunch and then outdoor sports for two hours before dismissal. If it rains, children will play indoor games after lunch. If the temperature is higher than 90F, tents will be set up outside before the sports activities.

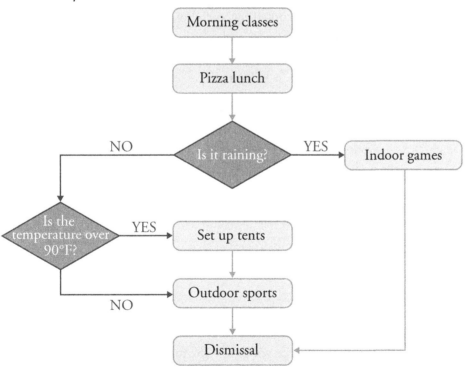

Paths, Circuits and Networks

Graph theory is a branch of mathematics that deals with the connectivity and symbolic representation of networks. Graphs can be used to represent a wide variety of types of information. They can be useful tools for solving problems that involve maps, hierarchies, directories, structures, communications networks and a range of other objects.

GRAPH: a set of points (or nodes) and lines (or edges) that connect some subset of these points

A GRAPH is a set of points (or nodes) and lines (or edges) that connect some subset of these points. A FINITE GRAPH has a limited number of both nodes and edges. An example graph is shown below. Note that not all of the nodes in a graph need be connected to other nodes.

FINITE GRAPH: a limited number of both nodes and edges

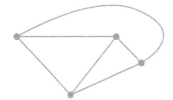

PATH: an uninterrupted succession of links in the same direction between points that join any two nodes

The edges of a graph may or may not have a specified direction or orientation; also, the edges (and nodes) may or may not have some assigned label or value. For instance, a graph representing airline flight paths might include nodes that represent cities and edges the represent the direction and distance of the paths between the cities.

CIRCUIT: a closed loop or path where the beginning node coincides with the terminal node

A PATH on a graph is an uninterrupted succession of links in the same direction between points that join any two nodes such as a flight path between two cities that passes through other cities. A closed loop or path where the beginning node coincides with the terminal node is known as a CIRCUIT.

TREE: a graph that does not include any closed loops

A TREE is a graph that does not include any closed loops. In addition, a tree has no unconnected nodes (separate nodes or groups of connected nodes constitute a separate tree—groups of several trees are called a forest). In addition, the edges that connect the nodes of a tree do not have a direction or orientation. A FINITE TREE, like a finite graph, has a limited number of edges and nodes. The graph shown above is not a tree, but it takes the form of a tree with a few alterations.

FINITE TREE: has a limited number of edges and nodes

NETWORK: a graph (directed or undirected) where each edge is assigned a positive real number in accordance with a specific function

A NETWORK is a graph (directed or undirected) where each edge is assigned a positive real number in accordance with a specific function. The function may correspond to the distance between two points on a map, for instance.

Example: Find the shortest path between points A and B on the following directed network.

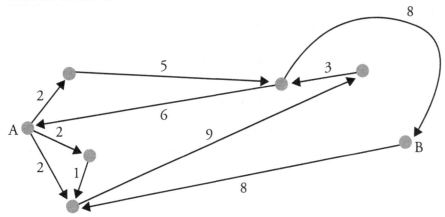

Taking careful note of the direction of the edges, find the possible routes through the graph from *A* to *B*. Choose the path with the smallest sum of the values along the associated edges.

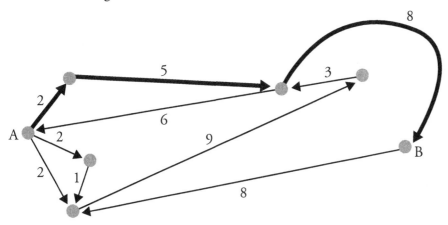

The total distance for the shortest path is 2 + 5 + 8 = 15.

Algorithms

An algorithm is a method of calculating; simply put, it can be multiplication, subtraction, or a combination of operations. When we work with computers and calculators, we employ algorithmic thinking, which means performing mathematical tasks by creating a sequential and often repetitive set of steps. A simple example would be to create an algorithm to generate the Fibonacci numbers utilizing the MR and M+ keys found on most calculators. The table below shows Entry made in the calculator, the value *x* seen in the display, and the value M contained in the memory.

ENTRY	ON/AC	1	M+	+	M+	MR	+	M+	MR	+	...
x	0	1	1	1	1	2	3	3	5	8	...
M	0	0	1	1	2	2	2	5	5	5	...

This eliminates the need to repeatedly enter required numbers.

Computers have to be programmed and many advanced calculators are programmable. A PROGRAM is the steps of an algorithm that are entered into a computer or calculator. The main advantage of using a program is that once the algorithm is entered, a result may be obtained by merely hitting a single keystroke to select the program, thereby eliminating the need to continually enter a large number of steps. Teachers find that programmable calculators are excellent for investigating "what if?" situations.

PROGRAM: the steps of an algorithm that are entered into a computer or calculator

Using graphing calculators or computer software has many advantages. The technology is better able to handle large data sets, such as the results of a science experiment and it is much easier to edit and sort the data and change the style of the graph to find its best representation. Furthermore, graphing calculators also provide a tool to plot statistics.

Concrete and visual representations can help demonstrate the logic behind operational algorithms. Blocks or other objects modeled on the base ten system are useful concrete tools. Base ten blocks represent ones, tens and hundreds. For example, modeling the partial sums algorithm with base ten blocks helps clarify the thought process. Consider the sum of 242 and 193. We represent 242 with 2 one hundred blocks, 4 ten blocks, and 2 one blocks. We represent 193 with 1 one hundred block, 9 ten blocks, and 3 one blocks. In the partial sums algorithm, we manipulate each place value separately and total the results. Thus, we group the hundred blocks, ten blocks, and one blocks and derive a total for each place value. We combine the place values to complete the sum.

An example of a visual representation of an operational algorithm is the modeling of a two-term multiplication as the area of a rectangle. For example, consider the product of 24 and 39. We can represent the product in geometric form. Note that the four sections of the rectangle equate to the four products of the partial products method.

	30	9
20	A = 600	A = 180
4	A = 120	A = 36

Thus, the final product is the sum of the areas or $600 + 180 + 120 + 36 = 936$.

SKILL 4B.7 Explore patterns in order to make conjectures, predictions, or generalizations

Even though arithmetic and geometric sequences are the most common patterns, one can have series based on other rules as well. In some problems, the student will not be given the rule that governs a pattern but will have to inspect the pattern to find out what the rule is.

Example: Find the next term in the series 1, 1, 2, 3, 5, 8, …

Inspecting the terms in the series, one finds that this pattern is neither arithmetic nor geometric. Every term in the series is a sum of the previous two terms. Thus, the next term = $5 + 8 = 13$.

This particular sequence is a well-known series named the Fibonacci sequence.

Just like the arithmetic and geometric sequences discussed before, other patterns can be created using algebraic variables. Patterns may also be pictorial. In each case, one can predict subsequent terms or find a missing term by first discovering the rule that governs the pattern.

Example: Find the next term in the sequence ax^2y, ax^4y^2, ax^6y^3, …

Inspecting the pattern we see that this is a geometric sequence with common ratio x^2y.

Thus, the next term = $ax^6y^3 \times x^2y = ax^8y^4$.

Example: Find the next term in the pattern:

Inspecting the pattern one observes that it has alternating squares and circles that include a number of hearts that increases by two for each subsequent term. Hence, the next term in the pattern will be as follows:

The following table represents the number of problems Mr. Rodgers is assigning his math students for homework each day, starting with the first day of class.

Day	1	2	3	4	5	6	7	8	9	10	11
Number of Problems	1	1	2	3	5	8	13				

If Mr. Rodgers continues this pattern, how many problems will he assign on the eleventh day?

If we look for a pattern, it appears that the number of problems assigned each day is equal to the sum of the problems assigned for the previous two days. We test this as follows:

Day 2 = 1 + 0 = 1
Day 3 = 1 + 1 = 2
Day 4 = 2 + 1 = 3
Day 5 = 3 + 2 = 5
Day 6 = 5 + 3 = 8
Day 7 = 8 + 5 = 13

Therefore, Day 8 would have 21 problems; Day 9, 34 problems; Day 10, 55 problems; and Day 11, 89 problems.

Suppose we have an equation $y = 2x + 1$. We construct a table of values in order to graph the equation to see if we can find a pattern.

X	-2	-1	0	1	2
Y	-3	-1	1	3	5

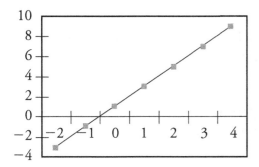

The pattern formed by the points is that they all lie on a line. We, therefore, can determine any solution of *y* by picking an *x*-coordinate and finding the corresponding point on the line. For example, if we want to know the solution of *y* when *x* is equal to 4, we find the corresponding point and see that *y* is equal to 9.

Example: Conjecture about pattern presented in tabular form.

Kepler discovered a relationship between the average distance of a planet from the sun and the time it takes the planet to orbit the sun. The following table shows the data for the six planets closest to the sun:

	MERCURY	VENUS	EARTH	MARS	JUPITER	SATURN
Average distance, *x*	0.387	0.723	1	1.523	5.203	9.541
*x*³	0.058	.378	1	3.533	140.852	868.524
Time, *y*	0.241	0.615	1	1.881	11.861	29.457
*y*²	0.058	0.378	1	3.538	140.683	867.715

Looking at the data in the table, we see that $x^3 = y^2$. We can conjecture the following function for Kepler's relationship: $y = \sqrt{x^3}$.

Representation of Patterns Using Symbolic Notation

Example: Find the recursive formula for the sequence 1, 3, 9, 27, 81…

We see that any term other than the first term is obtained by multiplying the preceding term by 3. Then, we may express the formula in symbolic notation as

$$a_n = 3a_{n-1}, a_1 = 1$$

where a represents a term, the subscript n denotes the place of the term in the sequence and the subscript $a - 1$ represents the preceding term.

Identification of Patterns of Change Created by Functions (e.g., Linear, Quadratic, Exponential)

A LINEAR FUNCTION is a function defined by the equation $f(x) = mx + b$.

> **LINEAR FUNCTION:** a function defined by the equation $f(x) = mx + b$

Example: A model for the distance traveled by a migrating monarch butterfly looks like f(t) = 80t, where t represents time in days. We interpret this to mean that the average speed of the butterfly is 80 miles per day and distance traveled may be computed by substituting the number of days traveled for t. In a linear function, there is a constant rate of change.

The standard form of a QUADRATIC FUNCTION is $f(x) = ax^2 + bx + c$.

> **QUADRATIC FUNCTION:** the standard form of a quadratic function is $f(x) = ax^2 + bx + c$

Example: What patterns appear in a table for y = x² − 5x + 6?

X	0	1	2	3	4	5
Y	6	2	0	0	2	6

We see that the values for y are symmetrically arranged.

An EXPONENTIAL FUNCTION is a function defined by the equation $y = ab^x$, where a is the starting value, b is the growth factor, and x tells how many times to multiply by the growth factor.

> **EXPONENTIAL FUNCTION:** a function defined by the equation $y = ab^x$, where a is the starting value, b is the growth factor, and x tells how many times to multiply by the growth factor

Example: y = 100(1.5)ˣ

X	0	1	2	3	4
Y	100	150	225	337.5	506.25

This is an exponential or multiplicative pattern of growth.

Iterative and Recursive Functional Relationships (e.g., Fibonacci Numbers)

The ITERATIVE PROCESS involves repeated use of the same steps. A RECURSIVE FUNCTION is an example of the iterative process. A recursive function is a function that requires the computation of all previous terms in order to find a subsequent term. Perhaps the most famous recursive function is the FIBONACCI SEQUENCE. This is the sequence of numbers 1, 1, 2, 3, 5, 8, 13, 21, 34, ... for which the next term is found by adding the previous two terms.

Sequences can be finite or infinite. A FINITE SEQUENCE is a sequence whose domain consists of the set {1, 2, 3, ... n} or the first n positive integers. An INFINITE SEQUENCE is a sequence whose domain consists of the set {1, 2, 3, ...}; which is in other words all positive integers.

A RECURRENCE RELATION is an equation that defines a sequence recursively; in other words, each term of the sequence is defined as a function of the preceding terms.

A real-life application would be using a recurrence relation to determine how much your savings would be in an account at the end of a certain period of time.

For example: You deposit $5,000 in your savings account. Your bank pays 5% interest compounded annually. How much will your account be worth at the end of 10 years?

Let V represent the amount of money in the account and V_n represent the amount of money after n years.

The amount in the account after n years equals the amount in the account after $n - 1$ years plus the interest for the nth year. This can be expressed as the recurrence relation V_0, where your initial deposit is represented by $V_0 = 5,000$.

$$V_0 = V_0$$
$$V_1 = 1.05V_0$$
$$V_2 = 1.05V_1 = (1.05)^2 V_0$$
$$V_3 = 1.05V_2 = (1.05)^3 V_0$$
$$......$$
$$V_n = 1.05V_{n-1} = (1.05)^n V_0$$

Inserting the values into the equation, you get $V_{10} = (1.05)^{10}(5,000) = 8,144$.

You determine that after investing $5,000 in an account earning 5% interest, compounded annually for 10 years, you would have $8,144.

ITERATIVE PROCESS: involves repeated use of the same steps

RECURSIVE FUNCTION: an example of the iterative process

FIBONACCI SEQUENCE: the sequence of numbers 1, 1, 2, 3, 5, 8, 13, 21, 34, ... for which the next term is found by adding the previous two terms

FINITE SEQUENCE: consists of the set {1, 2, 3, ... n} or the first n positive integers

INFINITE SEQUENCE: consists of the set {1, 2, 3, ...}; which is in other words all positive integers

RECURRENCE RELATION: an equation that defines a sequence recursively; each term of the sequence is defined as a function of the preceding terms

DOMAIN V
PROBLEM-SOLVING EXERCISES

Teachers can promote problem solving by allowing multiple attempts at problems. Teachers should be familiar with several specific problem-solving skills.

Guess-and-Check Method

The GUESS-AND-CHECK STRATEGY calls for students to make an initial guess at the solution, check the answer, and use the outcome to guide the next guess. With each successive guess, the student should get closer to the correct answer. Constructing a table from the guesses can help organize the data.

Example: There are 100 coins in a jar. Ten of them are dimes, and the rest are pennies and nickels. There are twice as many pennies as nickels. How many pennies and nickels are in the jar?

There are 90 total nickels and pennies in the jar (100 coins—10 dimes).

There are twice as many pennies as nickels. Make guesses that fulfill the criteria and adjust based on the answer found. Continue until we find the correct answer, 60 pennies, and 30 nickels.

NUMBER OF PENNIES	NUMBER OF NICKELS	TOTAL NUMBER OF PENNIES AND NICKELS
40	20	60
80	40	120
70	35	105
60	30	90

Working Backward

When solving a problem where the final result and the steps to reach the result are given, students must work backward to determine what the starting point must have been.

Example: Malcolm subtracted seven from his age and divided the result by 3. The final result was 4. What is Malcolm's age?

Work backward by reversing the operations:

$4 \times 3 = 12$;

$12 + 7 = 19$

Malcolm is 19 years old.

> **GUESS-AND-CHECK STRATEGY:** calls for students to make an initial guess at the solution, check the answer, and use the outcome to guide the next guess

Estimation and Testing for Reasonableness

Estimation and testing for reasonableness are related skills students should employ prior to and after solving a problem. These skills are particularly important when students use calculators to find answers.

Example: Find the sum of 4387 + 7226 + 5893.

$$4300 + 7200 + 5800 = 17300 \qquad \text{Estimation.}$$
$$4387 + 7226 + 5893 = 17506 \qquad \text{Actual sum.}$$

By comparing the estimate to the actual sum, students can determine that their answer is reasonable.

Breaking a Problem into Simpler Steps

Sometimes, the best way to solve a problem may be to break it into a series of simpler problems. This may be most appropriate in the following situations:

- A direct solution to the problem is too complicated

- The problem involves numbers that are either too small or too large

- You need to better understand the problem

- The computations are too complex

- The problem involves a diagram or a large array

Example: There are 20 people at a party. If each person shakes hands with all of the other guests, how many handshakes will there be?

We could take the long approach by seeing that the first person shakes hands with 19 other people, the second person has 18 other people with whom to shake hands, the third person has 17 other people with whom to shake hands, and then adding 1 through 19 handshakes.

A simpler approach would be to break the problem down into a smaller problem such as 4 people at a party. We see that the first person would shake hands with 3 other people, the second would shake hands with 2 other people, and the third person would shake hands with 1 other person. We can represent this as

$$(4 - 1) + (4 - 2) + (4 - 3) = 3 + 2 + 1 = 6$$

or

$$(n - 1) + (n - 2) + (n - 3) = 3n - 6 = 6$$

where n = the total number of people at the party. If we set the answer equal to x, we can solve for x as follows:

$$(n - 1)n - x = x$$
$$(20 - 1)(20) = 2x$$

$(19)(20) = 2x$
$380 = 2x$
$190 = x$

Using Partial Solutions

Example: A fish is 30 inches long. The head is as long as the tail. If the head was twice as long and the tail was its present length, the body would be 18 inches long. How long is the body?

Partial solution: Let x represent the head.

$2x + x + 18 = 30$
$3x = 12$
$x = 4$

We now create an equation to solve for the body of the fish with y representing the body.

$x + x + y = 30$
$2x + y = 30$
Substitute 4 for x.
$2(4) + y = 30$
$8 + y = 30$
$y = 22$

In this example, we are able to substitute the partial solution to solve for the variable in the problem's actual question.

Example: How many squares must be added to a 10-by-10 square to create an 11-by-11 square?

  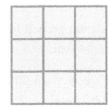

1-square   2-squares   3-squares

Partial solution: We determine that a 3-by-3 square has 5 more squares than a 2-by-2 square, which has 3 more squares than 1 square.

By examining the pattern, we see that we can answer the question by adding the dimension of the previous square (in this case, 10) to the dimension of the current square (in this case, 11) to answer the question. Twenty-one squares must be added to a 10-by-10 square to create an 11-by-11 square.

Exercise 1: Interpreting slope as a rate of change

Connection: Social Sciences/Geography

Real-Life Application: Slope is often used to describe a constant or average rate of change. These problems usually involve units of measure such as miles per hour or dollars per year.

Problem: The town of Verdant Slopes has been experiencing a boom in population growth. By the year 2000, the population had grown to 45,000, and by 2005, the population had reached 60,000.

Communicating about Algebra (a): Using the formula for slope as a model, find the average rate of change in population growth, expressing your answer in people per year.

Extension (b): Using the average rate of change determined in (a), predict the population of Verdant Slopes in the year 2010.

Solution (a): Let t represent the time and p represent population growth. The two observances are represented by (t_1, p_1) and (t_2, p_2).

First observance $= (t_1, p_1) = (2000, 45{,}000)$
Second observance $= (t_2, p_2) = (2005, 60{,}000)$

Use the formula for slope to find the average rate of change.

Rate of change $= \dfrac{p_2 - p_1}{t_2 - t_1}$

Substitute values.

$= \dfrac{60{,}000 - 45{,}000}{2005 - 2000}$

Simplify.

$= \dfrac{15{,}000}{5} = 3{,}000 \ people/year$

The average rate of change in population growth for Verdant Slopes between the years 2000 and 2005 was 3,000 people/year.

Solution (b): $3{,}000 \ people/year \times 5 \ years = 15{,}000 \ people$
$60{,}000 \ people + 15{,}000 \ people = 75{,}000 \ people$

At a continuing average rate of growth of 3,000 people/year, the population of Verdant Slopes could be expected to reach 75,000 by the year 2010.

Exercise 2

A. Find the midpoint between (5, 2) and (-13, 4).

Using the Midpoint Formula:

$\left(\dfrac{x_1 + x_2}{2}, \dfrac{y_1 + y_2}{2}\right) = \left(\dfrac{5 + (-13)}{2}, \dfrac{2 + 4}{2}\right) = \left(\dfrac{-8}{2}, \dfrac{6}{2}\right) = (-4, 3)$

B. Find the value of x_1 so that (-3, 5) is the midpoint between $(x_1, 6)$ and (-2, 4)

Using the Midpoint Formula:

$$(-3, 5) = (\frac{x_1 + x_2}{2}, \frac{y_1 + y_2}{2})$$
$$= (\frac{x_1 + (-2)}{2}, \frac{6 + 4}{2})$$
$$= (\frac{x_1 - 2}{2}, \frac{10}{2})$$
$$= (\frac{x_1 - 2}{2}, 5)$$

Separate out the x value to determine x_1.

$$-3 = \frac{x_1 - 2}{2}$$
$$-6 = x_1 - 2$$
$$-4 = x_1$$

C. Is $y = 3x - 6$ a bisector of the line segment with endpoints at (2, 4) and (8, -1)?

Find the midpoint of the line segment and then see if the midpoint is a point on the given line. Using the Midpoint Formula:

$$P = (\frac{2 + 8}{2}, \frac{4 + (-1)}{2}) = (\frac{10}{2}, \frac{4 - 1}{2}) = (5, \frac{3}{2}) = (5, 1.5)$$

Check to see if this point is on the line:

$$y = 3x - 6$$
$$y = 3(5) - 6 = 15 - 6 = 9$$

In order for this line to be a bisector, y must equal 1.5. However, since $y = 9$, the answer to the question is "No, this is not a bisector."

Exercise 3

A. One line passes through the points (-4, -6) and (4, 6); another line passes through the points (-5, -4) and (3, 8). Are these lines parallel, perpendicular, or neither?

Find the slopes.

$$m = \frac{y_2 - y_1}{x_2 - x_1}$$
$$m_1 = \frac{6 - (-6)}{4 - (-4)} = \frac{6 + 6}{4 + 4} = \frac{12}{8} = \frac{3}{2}$$
$$m_2 = \frac{8 - (-4)}{3 - (-5)} = \frac{8 + 4}{3 + 5} = \frac{12}{8} = \frac{3}{2}$$

Since the slopes are the same, the lines are parallel.

B. One line passes through the points (1, -3) and (0, -6); another line passes through the points (4, 1) and (-2, 3). Are these lines parallel, perpendicular, or neither?

Find the slopes.

$$m = \frac{y_2 - y_1}{x_2 - x_1}$$

$$m_1 = \frac{-6 - (-3)}{0 - 1} = \frac{-6 + 3}{-1} = \frac{-3}{-1} = 3$$

$$m_2 = \frac{3 - 1}{-2 - 4} = \frac{2}{-6} = \frac{1}{3}$$

The slopes are negative reciprocals, so the lines are perpendicular.

C. One line passes through the points (-2, 4) and (2, 5); another line passes through the points (-1, 0) and (5, 4). Are these lines parallel, perpendicular, or neither?

Find the slopes.

$$m = \frac{y_2 - y_1}{x_2 - x_1}$$

$$m_1 = \frac{5 - 4}{2 - (-2)} = \frac{1}{2 + 2} = \frac{1}{4}$$

$$m_2 = \frac{4 - 0}{5 - (-1)} = \frac{4}{5 + 1} = \frac{4}{6} = \frac{2}{3}$$

Since the slopes are not the same, the lines are not parallel. Since they are not negative reciprocals, they are not perpendicular either. Therefore, the answer is "neither."

Exercise 4

For 2000 through 2005, the consumption of a certain product sweetened with sugar, as a percent, $f(t)$, of the total consumption of the product, can be modeled by:

$f(t) = 75 + 37.25(0.615)^t$,

where $t = 2$ represents 2000.

A. Find a model for the consumption of the product sweetened with nonsugar sweeteners as a percent, $g(t)$, of the total consumption of the product.

Since 100% represents the total consumption of the product, the model can be found by subtracting the model for sugar-sweetened product from 100:

$$g(t) = 100 - (75 + 37.25(0.615)^t)$$

$$= 100 - 75 - 37.25(0.615)^t$$

$$= 25 - 37.25(0.615)^t$$

B. Sketch the graphs of f and g. Does the consumption of one type of product seem to be stabilizing compared to the other product? Explain.

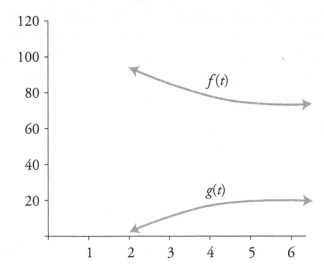

Yes, the consumption of the product sweetened with sugar (represented by $f(t)$) is decreasing less and less each year.

C. Sketch the graph of $f(x) = 2^x$. Does it have an x-intercept? What does this tell you about the number of solutions of the equation $2^x = 0$? Explain.

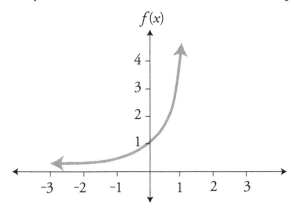

No, there is no solution. The solutions of $2^x = 0$ are the x-intercepts of $y = 2^x$.

Essential Tips for Every Math Teacher

Pedagogical principles and teaching methods are important for all teachers. They are particularly critical, though, for math teaching, since math teachers not only face the difficulty of communicating the subject matter to students but also that of surmounting an all-pervasive cultural fear of mathematics. Math teachers need to take particular care to foster learning in a nonthreatening environment that is at the same time genuinely stimulating and challenging.

The National Council of Teachers of Mathematics (NCTM) (*http://www.nctm.org/*) Principles and Standards emphasizes the teacher's obligation to support all students not only in developing basic mathematics knowledge and skills but also in their ability to understand and reason mathematically to solve problems relevant to today's world. The use of technology in the classroom is strongly advocated.

Resources for middle school teachers are available on the NCTM website at *http://www.nctm.org/resources/middle.aspx.*

The Mathematics Pathway (*http://msteacher.org/math.aspx*) on the National Science Digital Library (NSDL) Middle School Portal provides a very comprehensive and rich treasure trove of helpful material linking to various resources on the web including articles as well as interactive instructional modules on various topics.

The Drexel University Math Forum website provides the opportunity to interact with mentors and other math educators online. Some of the material on this website requires paid subscription but there are openly available archives as well. An overview of what the site provides is available at *http://mathforum.org/about_landing.html*. You may find the "Teacher2Teacher" service particularly useful; you can ask questions or browse the archives for a wealth of nitty-gritty everyday teaching information, suggestions and links to teaching tools.

Other instructional and professional development resources:

http://archives.math.utk.edu/k12.html

http://www.learnalberta.ca/Launch.aspx?content=/content/mesg/html/math6web/ math6shell.html

Pedagogical Principles

Maintaining a supportive, nonthreatening environment

Many students unfortunately perceive mathematics as a threat. This becomes a particular critical issue at the middle school level where they learn algebra for the first time and are required to think in new ways. Since fear "freezes" the brain and makes thinking really difficult, a student's belief that he is no good at math becomes a self-fulfilling prophecy. A teacher's primary task in this situation is to foster a learning environment where every student feels that he or she can learn to think mathematically. Here are some ways to go about this:

Accept all comments and questions

Acknowledge all questions and comments that students make. If what the student says is inaccurate or irrelevant to the topic in hand, point that out gently but also show your understanding of the thought process that led to the comment. This will encourage students to speak up in class and enhance their learning.

Set aside time for group work

Assign activities to groups of students comprised of mixed ability levels. It is often easier for students to put forward their own thoughts as part of a friendly group discussion than when they are sitting alone at their desks with a worksheet. The more proficient students can help the less able ones and at the same time clarify their own thinking. You will essentially be using the advanced students in the class as a resource in a manner that also contributes to their own development. The struggling students will feel supported by their peers and not isolated from them.

Encourage classroom discussion of math topics

For instance, let the whole class share different ways in which they approach a certain problem. It will give you insight into your students' ways of thinking and make it easier to help them. It will allow even those who just listen to understand and correct errors in their thinking without being put on the spot.

Engage and challenge students

Maintaining a nonthreatening environment should not mean "dumbing down" the math content in the classroom. The right level of challenge and relevance to their daily lives can help to keep students interested and learning. Here are some ideas:

Show connections to the real world

Use real life examples of math problems in your teaching. Some suggestions are given in the next section. Explain the importance of math literacy in society and the pitfalls of not being mathematically aware. An excellent reference is *The 10 Things All Future Mathematicians and Scientists Must Know* by Edward Zaccaro. The title of the book is misleading, since it deals with things that every educated person, not just mathematicians and scientists, should know.

Use technology

Use calculators and computers, including various online, interactive resources in your teaching. The natural affinity today's children have for these devices will definitely help them to engage more deeply in their math learning.

Demonstrate "messy" math

Children often have the mistaken belief that every math problem can be solved by following a particular set of rules; they either know the rules or they don't. In real life, however, math problems can be approached in different ways and often one has to negotiate several blind alleys before getting to the real solution. Children instinctively realize this while doing puzzles or playing games. They just don't associate this kind of thinking with classroom math. The most important insight any math teacher can convey to students is the realization that even if they don't know how to do a problem at first, they can think about it and figure it out as long as they are willing to stay with the problem and make mistakes in the process. An obvious way to do this, of course, is to introduce mathematical puzzles and games in the classroom. The best way, however, is for teachers themselves to take risks occasionally with unfamiliar problems and demonstrate to the class how one can work one's way out of a clueless state.

Show the reasoning behind rules

Even when it is not a required part of the curriculum, explain, whenever possible, how a mathematical rule is derived or how it is connected to other rules. For instance, in explaining the rule for finding the area of a trapezoid, show how one can get to it by thinking of the trapezoid as two triangles. This will reinforce the students' sense of mathematics as something that can be logically arrived at and not something for which they have to remember innumerable rules. Another way to reinforce this idea is to do the same problem using different approaches.

Be willing to take occasional side trips

Be flexible at times and go off topic in order to explore more deeply any questions or comments from the students. Grab a teaching opportunity even if it is irrelevant to the topic under discussion.

Help every student gain a firm grasp of fundamentals

While discussion, reasoning and divergent thinking is to be encouraged, it can only be done on a firm scaffolding of basic math knowledge. A firm grasp of math principles, for most people, does require rote exercises and doing more and more of the same problems. Just as practicing scales is essential for musical creativity, math creativity can only be built on a foundation strengthened by drilling and repetition. Many educators see independent reasoning and traditional rule-based drilling as opposing approaches. An effective teacher, however, must maintain a balance between the two and ensure that students have the basic tools they need to think independently.

Make sure all students actually know basic math rules and concepts

Test students regularly for basic math knowledge and provide reinforcement with additional practice wherever necessary.

Keep reviewing old material

Don't underestimate your students' ability to forget what they haven't seen in a while. Link new topics whenever possible with things your students have learned before and take the opportunity to review previous learning. Most math textbooks nowadays have a spiral review section created with this end in mind.

Keep mental math muscles strong

The calculator, without question, is a very valuable learning tool. Many students, unfortunately, use it as a crutch to the point that they lose the natural feel for numbers and ability to estimate that people develop through mental calculations. As a result, they are often unable to tell when they punch a wrong button and get a very unreasonable answer. Take your students through frequent mental calculation exercises; you can easily integrate it into class discussions. Teach them useful strategies for making mental estimates.

Specific Teaching Methods

Some commonly used teaching techniques and tools are described below, along with links to further information. The links provided in the first part of this chapter also provide a wealth of instructional ideas and material.

A very useful resource is the book *Family Math: The Middle School Years* from the Lawrence Hall of Science, University of California at Berkeley. Although this book was developed for use by families, teachers in school can choose from the many simple activities and games used to reinforce two significant middle school skills, algebraic reasoning and number sense. A further advantage is that all the activities are based on NCTM standards and each activity lists the specific math concepts that are covered.

Here are some tools you can use to make your teaching more effective:

- Classroom openers: To start off your class with stimulated, interested and focused students, provide a short opening activity every day. You can make use of thought-provoking questions, puzzles or tricks. Also use relevant puzzles or tricks to illustrate specific topics at any point in your class. The following website provides some ideas: *http://mathforum.org/k12/k12puzzles/*.

- Real-life examples: Connect math to other aspects of your students' lives by using examples and data from the real world whenever possible. It will not only keep them engaged, it will also help answer the perennial question "Why do we have to learn math?"

Here are some online resources to get you started:

1. Using weather concepts to teach math: *http://www.nssl.noaa.gov/edu/ideas/*

2. Election math in the classroom: *http://mathforum.org/t2t/faq/election.html*

3. Math worksheets related to the Iditarod, an annual Alaskan sled dog race: *http://www.educationworld.com/a_lesson/lesson/lesson302.shtml*

4. Graphing with real data: *http://www.middleweb.com/Graphing.html*

Manipulatives

Manipulatives can help all students learn; particularly those oriented more towards visual and kinesthetic learning. Here are some ideas for the use of manipulatives in the classroom:

1. Use tiles, pattern blocks or geoboards to demonstrate geometry concepts such as shapes, area and perimeter. In the example shown below, 12 tiles are used to form different rectangles.

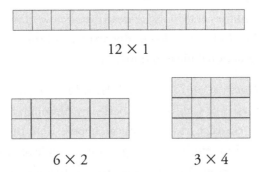

12×1

6×2 3×4

2. Stacks of blocks representing numbers are useful for teaching basic statistics concepts such as mean, median and mode. Rearranging the blocks to make each stack the same height would demonstrate the mean or average value of the data set. The example below shows a data set represented by stacks of blocks. Rearranging the blocks to make the height of each stack equal to three shows that this is the mean value.

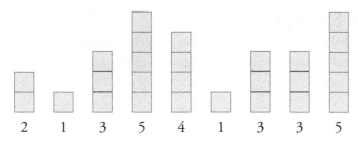

2 1 3 5 4 1 3 3 5

3. Tiles, blocks, or other countable manipulatives such as beans can also be used to demonstrate numbers in different bases. Each stack will represent a place with the number of blocks in the stack showing the place value.

4. Playing cards can be used for a discussion of probability.

5. Addition and subtraction of integers, positive and negative, is a major obstacle for many middle school students. Two sets of tiles, marked with pluses and minuses respectively, can be used to demonstrate these concepts visually with each "plus" tile canceling a "minus" tile.

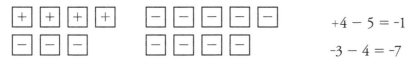

$$+4 - 5 = -1$$
$$-3 - 4 = -7$$

6. Percentages may be visualized using two parallel number lines, one showing the actual numbers and the other showing the percentages.

A practical demonstration of percent changes can be made by photocopying a figure using different copier magnifications.

7. Algeblocks are blocks designed specifically for the teaching of algebra with manipulatives: *http://www.etacuisenaire.com/algeblocks/algeblocks.jsp*

Software

Many of the online references in this section link to software for learning. A good site that provides easy to use virtual manipulatives as well as accompanying worksheets in some cases is the following: *http://boston.k12.ma.us/teach/technology/select/index.html*

Spreadsheets can be very effective math learning tools. Here are some ideas for using spreadsheets in the classroom: *http://www.angelfire.com/wi2/spreadsheet/necc.html*

Word Problem Strategies

Word problems, a challenge for many students even in elementary school, become more complicated and sometimes intimidating in the middle grades. Here are some ideas students can use to tackle them:

1. Identify significant words and numbers in the problem. Highlight or underline them. If necessary, write them in the form of a table.

2. Draw diagrams to clarify the problem. Put down the main items or events and numbers on the diagram and show the relationships between them.

3. Rewrite the problem using fewer and simpler words. One way is to have a standard format for this as shown in the example below.

   Problem: Calculate the cost of three pencils given that five pencils cost 25 cents.

   Rewrite as:
   Cost of 5 pencils = 25 cents
   Cost of 1 pencil = $\frac{25}{5}$ = 5 cents
   Cost of 3 pencils = $5 \times 3 = 15$ cents

4. If you have no idea how to approach the problem, try the guess and check approach at first. That will give you a sense of the kind of problem you are dealing with.

5. Create similar word problems of your own.

The Equation Rule

Solving algebraic equations is a challenge for many learners particularly when they think they need to remember many different rules. Emphasize the fact that they only need to keep only one rule in mind whether they are adding, subtracting, multiplying or dividing numbers or variables:

"Do the same thing to both sides."

A balance or teeter-totter metaphor can help to clarify their understanding of equations. You can also use manipulatives to demonstrate.

Mental Math Practice

Give students regular practice in doing mental math. The following website offers many mental calculation tips and strategies:

http://mathforum.org/k12/mathtips/mathtips.html

Because frequent calculator use tends to deprive students of a sense of numbers, they will often approach a sequence of multiplications and divisions the hard way. For instance, asked to calculate $770 \times 36/55$, they will first multiply 770 and 36 and then do a long division with the 55. They fail to recognize that both 770 and 55 can be divided by 11 and then by 5 to considerably simplify the problem. Give students plenty of practice in multiplying and dividing a sequence of integers and fractions so they are comfortable with canceling top and bottom terms.

Math Language

There is an explosion of new math words as students enter the middle grades and start learning algebra and geometry.

This website provides an animated, colorfully illustrated dictionary of math terms:

http://www.amathsdictionaryforkids.com/

Web Links

Algebra
Algebra in bite-size pieces with a quiz at the end:

http://library.thinkquest.org/20991/alg/index.html

Algebra II:

http://library.thinkquest.org/20991/alg2/index.html

Different levels plus a quiz:

http://www.math.com/homeworkhelp/Algebra.html

Clicking on the number leads to solution:

http://www.math.armstrong.edu/MathTutorial/index.html

Algebraic structures

Symbols and sets of numbers:

http://www.wtamu.edu/academic/anns/mps/math/mathlab/beg_algebra/beg_alg_tut2_sets.htm

Integers:

http://amby.com/educate/math/integer.html

Card game to add and subtract integers:

http://www.education-world.com/a_tsl/archives/03-1/lesson001.shtml

Multiplying integers:

http://www.aaastudy.com/mul65_x2.htm

Several complex number exercise pages:

http://math.about.com/od/complexnumbers/Complex_Numbers.htm

Polynomial equations and inequalities

Word problems system of equations:

http://regentsprep.org/regents/math/ALGEBRA/AE3/PracWord.htm

Inequality tutorial, examples, problems:

http://www.wtamu.edu/academic/anns/mps/math/mathlab/beg_algebra/beg_alg_tut18_ineq.htm

Graphing linear inequalities tutorial:

http://www.wtamu.edu/academic/anns/mps/math/mathlab/beg_algebra/beg_alg_tut24_ineq.htm

Quadratic equations tutorial, examples, problems:

http://www.wtamu.edu/academic/anns/mps/math/mathlab/col_algebra/col_alg_tut17_quad.htm

Synthetic division tutorial:

http://www.wtamu.edu/academic/anns/mps/math/mathlab/col_algebra/col_alg_tut37_syndiv.htm

Synthetic division Examples and problems:

http://www.tpub.com/math1/10h.htm

Functions
Function, domain, range intro and practice:

http://www.mathwarehouse.com/algebra/relation/math-function.php

Equations with rational expressions tutorial:

http://www.wtamu.edu/academic/anns/mps/math/mathlab/col_algebra/col_alg_tut15_rateq.htm

Practice simplifying radicals:

http://www.bhs87.org/math/practice/radicals/radicalpractice.htm

Radical equations—lesson and practice:

http://www.regentsprep.org/Regents/math/algtrig/ATE10/radlesson.htm

Logarithmic functions tutorial:

http://www.wtamu.edu/academic/anns/mps/math/mathlab/col_algebra/col_alg_tut43_logfun.htm

Linear algebra
Vector practice tip:

http://www.phy.mtu.edu/~suits/PH2100/vecdot.html

Geometry

http://library.thinkquest.org/20991/geo/index.html

http://www.math.com/students/homeworkhelp.html#geometry

http://regentsprep.org/Regents/math/geometry/math-GEOMETRY.htm

Parallelism
Parallel lines practice:

http://www.algebralab.org/lessons/lesson.aspx?file=Geometry_AnglesParallelLinesTransversals.xml

Plane Euclidean geometry
Geometry facts and practice:

http://www.aaaknow.com/geo.htm

Triangles intro and practice:

http://www.staff.vu.edu.au/mcaonline/units/geometry/triangles.html

Polygons exterior and interior angles practice:

http://regentsprep.org/regents/math/geometry/GG3/indexGG3.htm

Angles in circles practice:

http://regentsprep.org/Regents/math/geometry/GP15/PcirclesN2.htm

Congruence of triangles—lessons, practice:

http://regentsprep.org/Regents/math/geometry/GP4/indexGP4.htm

Pythagorean theorem and converse:

http://regentsprep.org/Regents/math/geometry/GP13/indexGP13.htm

Circle equation practice:

http://www.regentsprep.org/Regents/math/algtrig/ATC1/circlepractice.htm

Interactive parabola:

http://www.mathwarehouse.com/geometry/parabola/

Ellipse practice problems:

http://www.mathwarehouse.com/ellipse/equation-of-ellipse.php#equationOfEllipse

Three-dimensional geometry
3D figures intro and examples:

http://www.mathleague.com/help/geometry/3space.htm

Transformational geometry
Interactive transformational geometry practice on coordinate plane:

http://www.shodor.org/interactivate/activities/Transmographer/

Similar triangles practice:

http://regentsprep.org/regents/math/geometryGP11/indexGP11.htm

http://www.algebralab.org/practice/practice.aspx?file=Geometry_UsingSimilarTriangles.xml

Number theory
Natural Numbers:

http://online.math.uh.edu/MiddleSchool/Vocabulary/NumberTheoryVocab.pdf

GCF and LCM practice:

http://teachers.henrico.k12.va.us/math/ms/C1Files/01NumberSense/1_5/6035prac.htm

Probability and Statistics

Probability
Probability intro and practice:

http://www.mathgoodies.com/lessons/vol6/intro_probability.html

Permutation and combination practice:

http://www.regentsprep.org/Regents/math/algtrig/ATS5/PCPrac.htm

Statistics
Statistics lessons and interactive practice:

http://www.aaaknow.com/sta.htm

Range, mean, median, mode exercises:

http://www.mathgoodies.com/lessons/toc_vol8.html

http://regentsprep.org/REgents/math/ALGEBRA/AD2/Pmeasure.htm

SAMPLE TEST

SAMPLE TEST

Directions: Read each item and select the best response.

(Average)(Skill 1.1)

1. What would be the total cost of a suit for $295.99 and a pair of shoes for $69.95 including 6.5% sales tax?

   A. $389.73

   B. $398.37

   C. $237.86

   D. $315.23

(Average) (Skill 1.5)

2. A student had 60 days to appeal the results of an exam. If the results were received on March 23, what was the last day that the student could appeal?

   A. May 21

   B. May 22

   C. May 23

   D. May 24

(Average) (Skill 1.1)

3. Evaluate: $\frac{1}{3} - \frac{1}{2} + \frac{1}{6}$

   A. $\frac{5}{6}$

   B. $\frac{2}{3}$

   C. 0

   D. 1

(Average) (Skill 1.19)

4. Solve for x: $3(5 + 3x) - 8 = 88$

   A. 30

   B. 9

   C. 4.5

   D. 27

(Easy) (Skill 1.5)

5. Sandra has $34.00, Carl has $42.00. How much more does Carl have than Sandra? Which would be the best method for finding the answer?

   A. Addition

   B. Subtraction

   C. Division

   D. Both A and B are equally correct

(Easy) (Skill 1.14)

6. Express in symbols: "x is greater than seven and less than or equal to fifteen."

   A. $7 < x \le 15$

   B. $7 > x \ge 15$

   C. $15 \le x < 7$

   D. $7 < x = 15$

(Easy) (Skill 1.6)

7. Which of the following is correct?

   A. $2365 > 2340$

   B. $0.75 > 1.25$

   C. $\frac{3}{4} < \frac{1}{16}$

   D. $-5 < -6$

(Average) (Skill 1.11)

8. Given that n is a positive even integer, $5n + 4$ will always be divisible by:

   A. 4

   B. 5

   C. $5n$

   D. 2

(Easy) (Skill 1.2)

9. A student turns in a paper with this type of error:

   $7 + 16 \div 8 \times 2 = 8$

   $8 - 3 \times 3 + 4 = -5$

   In order to remediate this error, a teacher should:

   A. Review and drill basic number facts

   B. Emphasize the importance of using parentheses in simplifying expressions

   C. Emphasize the importance of working from left to right when applying the order of operations

   D. Do nothing; these answers are correct

(Easy) (Skill 1.7)

10. Given W = whole numbers

    N = natural numbers

    Z = integers

    R = rational numbers

    I = irrational numbers

    which of the following is not true?

    A. $R \subset I$

    B. $W \subset Z$

    C. $Z \subset R$

    D. $N \subset W$

(Easy) (Skill 1.7)

11. Which of the following is an irrational number?

    A. .362626262...

    B. $4\frac{1}{3}$

    C. $\sqrt{5}$

    D. $-\sqrt{16}$

(Rigorous) (Skill 1.7)

12. Which denotes a complex number?

    A. 3.678678678...

    B. $-\sqrt{27}$

    C. $123^{\frac{1}{2}}$

    D. $(-100)^{\frac{1}{2}}$

(Rigorous) (Skill 1.7)

13. Choose the correct statement:

    A. Rational and irrational numbers are both proper subsets of the set of real numbers

    B. The set of whole numbers is a proper subset of the set of natural numbers

    C. The set of integers is a proper subset of the set of irrational numbers

    D. The set of real numbers is a proper subset of the natural, whole, integers, rational, and irrational numbers

(Easy) (Skill 1.6)

14. How many real numbers lie between -1 and $+1$?

    A. 0

    B. 1

    C. 17

    D. An infinite number

(Average) (Skill 1.6)

15. Choose the set in which the members are not equivalent.

    A. $\frac{1}{2}$, 0.5, 50%

    B. $\frac{10}{5}$, 2.0, 200%

    C. $\frac{3}{8}$, 0.385, 38.5%

    D. $\frac{7}{10}$, 0.7, 70%

(Average) (Skill 1.6)

16. Change $.\overline{63}$ into a fraction in simplest form.

    A. $\frac{63}{100}$

    B. $\frac{7}{11}$

    C. $6\frac{3}{10}$

    D. $\frac{2}{3}$

(Rigorous) (Skill 1.11)

17. Which of the following is always composite if x is odd, y is even, and both x and y are greater than or equal to 2?

    A. $x + y$

    B. $3x + 2y$

    C. $5xy$

    D. $5x + 3y$

(Average) (Skill 1.8)

18. Express .0000456 in scientific notation.

    A. 4.56×10^{-4}

    B. 45.6×10^{-6}

    C. 4.56×10^{-6}

    D. 4.56×10^{-5}

(Easy) (Skill 1.5)

19. Mr. Brown feeds his cat premium cat food which costs $40 per month. Approximately how much will it cost to feed her for one year?

    A. $500

    B. $400

    C. $80

    D. $4800

(Average) (Skill 1.8)

20. $(3.8 \times 10^{17}) \times (.5 \times 10^{-12})$

    A. 19×10^{5}

    B. 1.9×10^{5}

    C. 1.9×10^{6}

    D. 1.9×10^{7}

(Average) (Skill 1.8)

21. 2^{-3} is equivalent to

    A. 0.8

    B. −0.8

    C. 125

    D. 0.125

(Rigorous) (Skill 1.8)

22. $\dfrac{3.5 \times 10^{-10}}{0.7 \times 10^{4}}$

    A. 0.5×10^{6}

    B. 5.0×10^{-6}

    C. 5.0×10^{-14}

    D. 0.5×10^{-14}

(Rigorous) (Skill 1.17)

23. Simplify $\dfrac{\frac{3}{4}x^2y^{-3}}{\frac{2}{3}xy}$

    A. $\dfrac{1}{2}xy^{-4}$

    B. $\dfrac{1}{2}x^{-1}y^{-4}$

    C. $\dfrac{9}{8}xy^{-4}$

    D. $\dfrac{9}{8}xy^{-2}$

(Average) (Skill 1.17)

24. Which of the following is incorrect?

    A. $(x^2y^3)^2 = x^4y^6$

    B. $m^2(2n)^3 = 8m^2n^3$

    C. $\dfrac{(m^3n^4)}{(m^2n^2)} = mn^2$

    D. $(x + y^2)^2 = x^2 + y^4$

(Rigorous) (Skill 1.8)

25. Evaluate: $3^{\frac{1}{2}}(9^{\frac{1}{3}})$

    A. $27^{\frac{5}{6}}$

    B. $9^{\frac{7}{12}}$

    C. $3^{\frac{5}{6}}$

    D. $3^{\frac{6}{7}}$

(Rigorous) (Skill 1.9)

26. Solve: $\sqrt{75} + \sqrt{147} - \sqrt{48}$

    A. 174

    B. $12\sqrt{3}$

    C. $8\sqrt{3}$

    D. 74

(Average) (Skill 1.9)

27. Simplify: $\sqrt{27} + \sqrt{75}$

    A. $8\sqrt{3}$

    B. 34

    C. $34\sqrt{3}$

    D. $15\sqrt{3}$

(Average) (Skill 1.11)

28. Given that x, y, and z are prime numbers, which of the following is true?

    A. $x + y$ is always prime

    B. xyz is always prime

    C. xy is sometimes prime

    D. $x + y$ is sometimes prime

(Average) (Skill 1.11)

29. Find the GCF of $2^2 \times 3^2 \times 5$ and $2^2 \times 3 \times 7$.

    A. $2^5 \times 3^3 \times 5 \times 7$

    B. $2 \times 3 \times 5 \times 7$

    C. $2^2 \times 3$

    D. $2^3 \times 3^2 \times 5 \times 7$

(Average) (Skill 1.11)

30. Given even numbers x and y, which could be the LCM of x and y?

    A. $\dfrac{xy}{2}$

    B. $2xy$

    C. $4xy$

    D. xy

(Average) (Skill 1.2)

31. $24 - 3 \times 7 + 2 =$

   A. 5

   B. 149

   C. -3

   D. 189

(Average) (Skill 1.2)

32. $7t - 4 \times 2t + 3t \times 4 \div 2 =$

   A. $5t$

   B. 0

   C. $31t$

   D. $18t$

(Rigorous) (Skill 1.12)

33. Joe reads 20 words/min., and Jan reads 80 words/min. How many minutes will it take Joe to read the same number of words that it takes Jan 40 minutes to read?

   A. 10

   B. 20

   C. 80

   D. 160

(Easy) (Skill 1.12)

34. If three cups of concentrate are needed to make 2 gallons of fruit punch, how many cups are needed to make 5 gallons?

   A. 6 cups

   B. 7 cups

   C. 7.5 cups

   D. 10 cups

(Average) (Skill 1.12)

35. A sofa sells for $520. If the retailer makes a 30% profit, what was the wholesale price?

   A. $400

   B. $676

   C. $490

   D. $364

(Rigorous) (Skill 1.16)

36. Simplify: $\dfrac{10}{1 + 3i}$

   A. $-1.25(1 - 3i)$

   B. $1.25(1 + 3i)$

   C. $1 + 3i$

   D. $1 - 3i$

(Easy) (Skill 1.3)

37. Which statement is an example of the identity axiom of addition?

   A. $3 + -3 = 0$

   B. $3x = 3x + 0$

   C. $3 \times \frac{1}{3} = 1$

   D. $3 + 2x = 2x + 3$

(Average) (Skill 1.3)

38. **Which axiom is incorrectly applied?**

$3x + 4 = 7$

Step a. $3x + 4 - 4 = 7 - 4$
additive equality

Step b. $3x + 4 - 4 = 3$
commutative axiom of addition

Step c. $3x + 0 = 3$
additive inverse

Step d. $3x = 3$
additive identity

A. Step a

B. Step b

C. Step c

D. Step d

(Average) (Skill 1.3)

39. **Which of the following sets is closed under division?**

A. Integers

B. Rational numbers

C. Natural numbers

D. Whole numbers

(Average) (Skill 1.4)

40. **Which of the following does not correctly relate an inverse operation?**

A. $a - b = a + -b$

B. $a \times b = b \div a$

C. $\sqrt{a^2} = a$

D. $a \times \frac{1}{a} = 1$

(Easy) (Skill 1.3)

41. **Given a, b, y, and z are real numbers and $ay + b = z$, prove $y = \dfrac{z + (-b)}{a}$**

STATEMENT	REASON
1. $ay + b = z$	Given
2. $-b$ is a real number	Closure
3. $(ay + b) + -b = z + -b$	Identity property of addition
4. $ay + (b + -b) = z + -b$	Associative
5. $ay + 0 = z + -b$	Additive inverse
6. $ay = z + -b$	Identity property of addition
7. $y = \dfrac{z + (-b)}{a}$	Division

Which reason is incorrect for the corresponding statement?

A. Step 3

B. Step 4

C. Step 5

D. Step 6

(Easy) (Skill 1.3)

42. **Which of the following sets is closed under division?**

I. $\{\frac{1}{2}, 1, 2, 4\}$

II. $\{-1, 1\}$

III. $\{-1, 0, 1\}$

A. I only

B. II only

C. III only

D. I and II

(Easy) (Skill 1.4)

43. Which of the following illustrates an inverse property?

    A. $a + b = a - b$

    B. $a + b = b + a$

    C. $a + 0 = a$

    D. $a + (-a) = 0$

(Rigorous) (Skill 1.19)

44. A boat travels 30 miles upstream in three hours. It makes the return trip in one and a half hours. What is the speed of the boat in still water?

    A. 10 mph

    B. 15 mph

    C. 20 mph

    D. 30 mph

(Easy) (Skill 2.5)

45. Given $XY \cong YZ$ and $\angle AYX \cong \angle AYZ$. Prove $\triangle AYZ \cong \triangle AYX$.

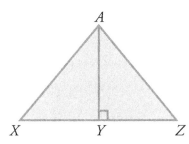

    1. $XY \cong YZ$

    2. $\angle AYX \cong \angle AYZ$

    3. $AY \cong AY$

    4. $\triangle AYZ \cong \triangle AYX$

    Which property justifies step 3?

    A. Reflexive

    B. Symmetric

    C. Transitive

    D. Identity

(Average) (Skill 2.6)

46. Given $l_1 \parallel l_2$ prove $\angle b \cong \angle e$.

    1. $\angle b \cong \angle d$ vertical angle theorem

    2. $\angle d \cong \angle e$ alternate interior angle theorem

    3. $\angle b \cong \angle e$ symmetric axiom of equality

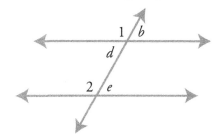

    Which step is incorrectly justified?

    A. Step 1

    B. Step 2

    C. Step 3

    D. No error

(Easy) (Skill 2.8)

47. Which of the following shapes is not a parallelogram?

    I
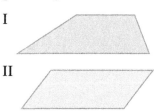

    II

    III

    A. I & III

    B. II & III

    C. I

    D. I, II & III

(Easy) (Skill 2.1)

48. 4 lbs 4 oz is equal to:

   A. 20 oz

   B. 4.25 lb

   C. 4.33 oz

   D. 64 oz

(Rigorous) (Skill 2.5)

49. Given similar polygons with corresponding sides of lengths 9 and 15, find the perimeter of the smaller polygon if the perimeter of the larger polygon is 150 units.

   A. 54

   B. 135

   C. 90

   D. 126

(Average) (Skill 2.6)

50. Given $l_1 \parallel l_2$ which of the following is true?

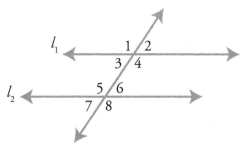

   A. ∠1 and ∠8 are congruent and alternate interior angles

   B. ∠2 and ∠3 are congruent and corresponding angles

   C. ∠3 and ∠4 are adjacent and supplementary angles

   D. ∠3 and ∠5 are adjacent and supplementary angles

(Rigorous) (Skill 2.9)

51. Given the regular hexagon below, determine the measure of angle 1.

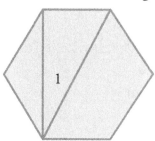

   A. 30°

   B. 60°

   C. 120°

   D. 45°

(Average) (Skill 2.5)

52. Given $QS \cong TS$ and $RS \cong US$, prove $\triangle QRS \cong \triangle TUS$.

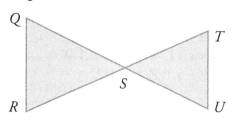

1. $QS \cong TS$	1. Given
2. $RS \cong US$	2. Given
3. $\angle TSU \cong \angle QSR$	3. ?
4. $\triangle TSU \cong \triangle QSR$	4. SAS

Give the reason which justifies step 3.

   A. Congruent parts of congruent triangles are congruent

   B. Reflexive axiom of equality

   C. Alternate interior angle theorem

   D. Vertical angle theorem

(Rigorous) (Skill 2.7)

53. In the figure below, what is the value of *x*?

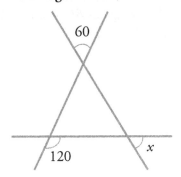

Note: Figure not drawn to scale.

A. 50

B. 60

C. 75

D. 80

(Rigorous) (Skill 2.12)

54. Line *p* has a negative slope and passes through the point (0, 0). If line *q* is perpendicular to line *p*, which of the following must be true?

A. Line *q* has a negative *y*-intercept

B. Line *q* passes through the point (0, 0)

C. Line *q* has a positive slope

D. Line *q* has a positive *y*-intercept

(Rigorous) (Skill 2.12)

55. What is the slope of any line parallel to the line $2x + 4y = 4$?

A. –2

B. –1

C. $-\frac{1}{2}$

D. 2

(Average) (Skill 2.5)

56. Prove $\triangle HYM \cong \triangle KZL$, given $XZ \cong XY$, $\angle L \cong \angle M$ and $YL \cong MZ$

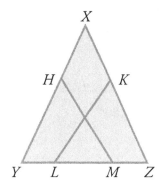

1. $XZ \cong XY$	1. Given
2. $\angle Y \cong \angle Z$	2. ?
3. $\angle L \cong \angle M$	3. Given
4. $YL \cong MZ$	4. Given
5. $LM \cong LM$	5. ?
6. $YM \cong LZ$	6. Add
7. $\triangle HYM \cong \triangle KZL$	7. ASA

Which could be used to justify steps 3 and 5?

A. CPCTC, Identity

B. Isosceles Triangle Theorem, Identity

C. SAS, Reflexive

D. Isosceles Triangle Theorem, Reflexive

(Rigorous) (Skill 2.9)

57. What is the degree measure of an interior angle of a regular 10-sided polygon?

A. 18°

B. 36°

C. 144°

D. 54°

(Rigorous) (Skill 2.7)

58. Which of the following statements is true about the number of degrees in each angle?

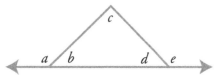

A. $a + b + c = 180°$

B. $a = e$

C. $b + c = e$

D. $c + d = e$

(Average) (Skill 2.5)

59. What method could be used to prove the below triangles congruent?

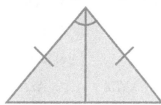

A. SSS

B. SAS

C. AAS

D. SSA

(Average) (Skill 2.5)

60. Which postulate could be used to prove $\triangle ABD \cong \triangle CEF$, given $BC \cong DE$, $\angle C \cong \angle D$, and $AD \cong CF$?

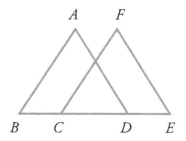

A. ASA

B. SAS

C. SAA

D. SSS

(Rigorous) (Skill 2.5)

61. Which theorem can be used to prove $\triangle BAK \cong \triangle MKA$?

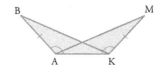

A. SSS

B. ASA

C. SAS

D. AAS

(Average) (Skill 2.5)

62. Given similar polygons with corresponding sides 6 and 8, what is the area of the smaller if the area of the larger is 64?

A. 48

B. 36

C. 144

D. 78

(Average) (Skill 2.5)

63. In similar polygons, if the perimeters are in a ratio of $x : y$, the sides are in a ratio of:

    A. $x : y$

    B. $x^2 : y^2$

    C. $2x : y$

    D. $\frac{1}{2}x : y$

(Rigorous) (Skill 2.7)

64. ALM is a right triangle with the right angle at A. Given altitude AK with measurements as indicated, determine the length of AK.

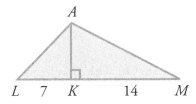

    A. 98

    B. $7\sqrt{2}$

    C. $\sqrt{21}$

    D. $7\sqrt{3}$

(Easy) (Skill 2.2)

65. The following diagram is most likely used in deriving a formula for which of the following?

    A. The area of a rectangle

    B. The area of a triangle

    C. The perimeter of a triangle

    D. The surface area of a prism

(Rigorous) (Skill 2.2)

66. Compute the area of the shaded region, given a radius of 5 meters. O is the center.

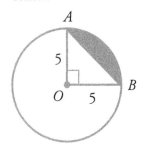

    A. 7.13 cm²

    B. 7.13 m²

    C. 78.5 m²

    D. 19.63 m²

(Rigorous) (Skill 2.3)

67. Ginny and Nick head back to their respective colleges after being home for the weekend. They leave their house at the same time and drive for 4 hours. Ginny drives due south at the average rate of 60 miles per hour and Nick drives due east at the average rate of 60 miles per hour. What is the straight-line distance between them, in miles, at the end of the 4 hours?

    A. $120\sqrt{2}$

    B. 240

    C. $240\sqrt{2}$

    D. 288

(Average) (Skill 2.3)

68. If $AC = 12$, determine BC.

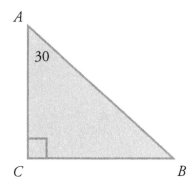

A. 6

B. 4

C. $6\sqrt{3}$

D. $3\sqrt{6}$

(Easy) (Skill 2.8)

69. Given that $QO \perp NP$ and $QO = NP$, quadrilateral $NOPQ$ can most accurately be described as a

A. Parallelogram

B. Rectangle

C. Square

D. Rhombus

(Average) (Skill 2.7)

70. Choose the correct statement concerning the median and altitude in a triangle.

A. The median and altitude of a triangle may be the same segment

B. The median and altitude of a triangle are always different segments

C. The median and altitude of a right triangle is always the same segment

D. The median and altitude of an isosceles triangle is always the same segment

(Average) (Skill 2.12)

71. Find the distance between (3, 7) and (-3, 4).

A. 9

B. 45

C. $3\sqrt{5}$

D. $5\sqrt{3}$

(Average) (Skill 2.12)

72. Find the midpoint of (2, 5) and (7, -4).

A. (9, -1)

B. (5, 9)

C. $\left(\frac{9}{2}, -\frac{1}{2}\right)$

D. $\left(\frac{9}{2}, \frac{1}{2}\right)$

(Rigorous) (Skill 2.12)

73. Given segment AC with B as its midpoint find the coordinates of C if $A = (5, 7)$ and $B = (3, 6.5)$.

A. (4, 6.5)

B. (1, 6)

C. (2, 0.5)

D. (16, 1)

(Average) (Skill 2.10)

74. What is the measure of major arc AL?

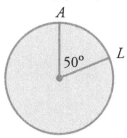

A. 50°

B. 25°

C. 100°

D. 310°

(Rigorous) (Skill 2.10)

75. If arc *KR* = 70° what is the measure of ∠*M*?

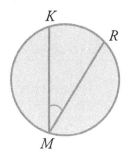

A. 290°

B. 35°

C. 140°

D. 110°

(Easy) (Skill 2.6)

76. The following construction can be completed to make

A. An angle bisector

B. Parallel lines

C. A perpendicular bisector

D. Skew lines

(Easy) (Skill 2.7)

77. A line from *R* to *K* will form

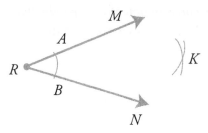

A. An altitude of *RMN*

B. A perpendicular bisector of *MN*

C. A bisector of *MRN*

D. A vertical angle

(Easy) (Skill 2.4)

78. An isosceles triangle has:

A. Three equal sides

B. Two equal sides

C. No equal sides

D. Two equal sides in some cases, no equal sides in others

(Rigorous) (Skill 2.10)

79. What is the measure of minor arc *AD*, given measure of arc *PS* is 40° and *m* < *K* = 10?

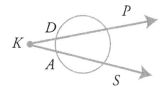

A. 50°

B. 20°

C. 30°

D. 25°

(Average) (Skill 2.6)

80. Choose the diagram which illustrates the construction of a perpendicular to the line at a given point on the line.

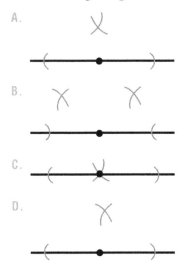

A.

B.

C.

D.

(Easy) (Skill 2.1)

81. Seventh grade students are working on a project using nonstandard measurement. Which would not be an appropriate instrument for measuring the length of the classroom?

A. A student's foot

B. A student's arm span

C. A student's jump

D. All are appropriate

(Average) (Skill 2.1)

82. The speed of light in space is about 3×10^8 meters per second. Express this in Kilometers per hour.

A. 1.08×10^9 Km/hr

B. 3.0×10^{11} Km/hr

C. 1.08×10^{12} Km/hr

D. 1.08×10^{15} Km/hr

(Easy) (Skill 2.1)

83. The mass of a Chips Ahoy cookie would be close to:

A. 1 kilogram

B. 1 gram

C. 15 grams

D. 15 milligrams

(Average) (Skill 2.13)

84. A man's waist measures 90 cm. What is the greatest possible error for the measurement?

A. ± 1 m

B. ± 8 cm

C. ± 1 cm

D. ± 5 mm

(Easy) (Skill 2.1)

85. 3 km is equivalent to:

A. 300 cm

B. 300 m

C. 3000 cm

D. 3000 m

(Average) (Skill 2.1)

86. 4 square yards is equivalent to:

A. 12 square feet

B. 48 square feet

C. 36 square feet

D. 108 square feet

(Rigorous) (Skill 2.2)

87. If a circle has an area of 25 cm², what is its circumference to the nearest tenth of a centimeter?

   A. 78.5 cm

   B. 17.7 cm

   C. 8.9 cm

   D. 15.7 cm

(Rigorous) (Skill 2.2)

88. Find the area of the figure below.

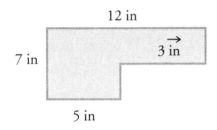

   A. 56 in²

   B. 27 in²

   C. 71 in²

   D. 170 in²

(Rigorous) (Skill 2.2)

89. Find the area of the shaded region given square *ABCD* with side *AB* = 10m and circle *E*.

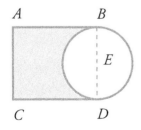

   A. 178.5 m²

   B. 139.25 m²

   C. 71 m²

   D. 60.75 m²

(Rigorous) (Skill 2.2)

90. Compute the area of the polygon shown below.

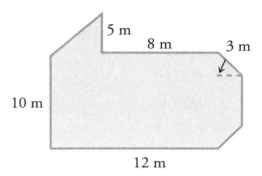

   A. 178 m²

   B. 154 m²

   C. 43 m²

   D. 188 m²

(Rigorous) (Skill 2.2)

91. Find the area of the figure pictured below.

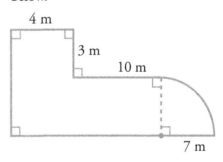

   A. 136.47 m²

   B. 148.48 m²

   C. 293.86 m²

   D. 178.47 m²

(Rigorous) (Skill 2.2)

92. Given a 30 meter × 60 meter garden with a circular fountain with a 5 meter radius, calculate the area of the portion of the garden not occupied by the fountain.

    A. 1721 m²

    B. 1879 m²

    C. 2585 m²

    D. 1015 m²

(Rigorous) (Skill 2.2)

93. Determine the area of the shaded region of the trapezoid in terms of *x* and *y*.

    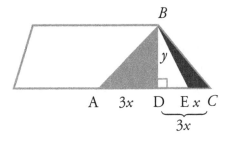

    A. 4*xy*

    B. 2*xy*

    C. 3*x*²*y*

    D. There is not enough information given

(Rigorous) (Skill 2.2)

94. If the radius of a right circular cylinder is doubled, how does its volume change?

    A. No change

    B. Also is doubled

    C. Four times the original

    D. Pi times the original

(Rigorous) (Skill 2.2)

95. Determine the volume of a sphere to the nearest cm if the surface area is 113 cm².

    A. 113 cm³

    B. 339 cm³

    C. 37.7 cm³

    D. 226 cm³

(Rigorous) (Skill 2.2)

96. Compute the surface area of the prism.

    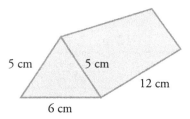

    A. 204

    B. 216

    C. 360

    D. 180

(Rigorous) (Skill 2.2)

97. If the base of a regular square pyramid is tripled, how does its volume change?

    A. Double the original

    B. Triple the original

    C. Nine times the original

    D. No change

(Easy) (Skill 2.2)

98. **How does lateral area differ from total surface area in prisms, pyramids, and cones?**

   A. For the lateral area, only use surfaces perpendicular to the base

   B. They are both the same

   C. The lateral area does not include the base

   D. The lateral area is always a factor of pi

(Average) (Skill 2.2)

99. **If the area of the base of a cone is tripled, the volume will be**

   A. The same as the original

   B. 9 times the original

   C. 3 times the original

   D. 3 pi times the original

(Rigorous) (Skill 2.2)

100. **Find the height of a box with surface area of 94 sq. ft. with a width of 3 feet and a depth of 4 feet.**

   A. 3 ft.

   B. 4 ft.

   C. 5 ft

   D. 6 ft.

(Rigorous) (Skill 3.1)

101. $f(x) = 3x - 2; f^{-1}(x) =$

   A. $3x + 2$

   B. $\frac{x}{6}$

   C. $2x - 3$

   D. $\frac{(x + 2)}{3}$

(Average) (Skill 1.19)

102. **Solve for x: $3x + 5 \geq 8 + 7x$**

   A. $x \geq -\frac{3}{4}$

   B. $x \leq -\frac{3}{4}$

   C. $x \geq \frac{3}{4}$

   D. $x \leq \frac{3}{4}$

(Rigorous) (Skill 1.21)

103. **Solve for x: $\left| 2x + 3 \right| > 4$**

   A. $-\frac{7}{2} > x > \frac{1}{2}$

   B. $-\frac{1}{2} > x > \frac{7}{2}$

   C. $x < \frac{7}{2}$ or $x > -\frac{1}{2}$

   D. $x < -\frac{7}{2}$ or $x > \frac{1}{2}$

(Rigorous) (Skill 1.19)

104. **Graph the solution: $\left| x \right| + 7 < 13$**

(Rigorous) (Skill 1.14)

105. **Solve for $v_0 : d = at(v_t - v_0)$**

   A. $v_0 = atd - v_t$

   B. $v_0 = d - atv_t$

   C. $v_0 = atv_t - d$

   D. $v_0 = \frac{(atv_t - d)}{at}$

(Average) (Skill 1.21)

106. Solve for x: $18 = 4 + |2x|$

   A. $\{-11, 7\}$

   B. $\{-7, 0, 7\}$

   C. $\{-7, 7\}$

   D. $\{-11, 11\}$

(Rigorous) (Skill 3.3)

107. Which graph represents the solution set for $x^2 - 5x > -6$?

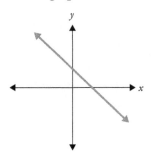

(Average) (Skill 3.3)

108. Which equation is represented by the below graph?

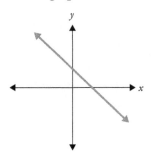

   A. $x - y = 3$

   B. $x - y = -3$

   C. $x + y = 3$

   D. $x + y = -3$

(Easy) (Skill 3.4)

109. Identify the proper sequencing of sub-skills when teaching graphing inequalities in two dimensions.

   A. Shading regions, graphing lines, graphing points, determining whether a line is solid or broken

   B. Graphing points, graphing lines, determining whether a line is solid or broken, shading regions

   C. Graphing points, shading regions, determining whether a line is solid or broken, graphing lines

   D. Graphing lines, determining whether a line is solid or broken, graphing points, shading regions

(Rigorous) (Skill 3.3)

110. What is the equation of the below graph?

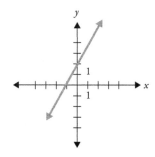

   A. $2x + y = 2$

   B. $2x - y = -2$

   C. $2x - y = 2$

   D. $2x + y = -2$

(Rigorous) (Skill 3.3)

111. Which graph represents the equation of
$y = x^2 + 3x$?

A.

B.

C.

D.

(Rigorous) (Skill 1.20)

112. Solve for x. $3x^2 - 2 + 4(x^2 - 3) = 0$

A. $\{-\sqrt{2}, \sqrt{2}\}$

B. $\{2, -2\}$

C. $\{0, \sqrt{3}, -\sqrt{3}\}$

D. $\{7, -7\}$

(Average) (Skill 1.14)

113. Which of the following is a factor of
$6 + 48m^3$?

A. $(1 + 2m)$

B. $(1 - 8m)$

C. $(1 + m - 2m)$

D. $(1 - m + 2m)$

(Average) (Skill 1.14)

114. Factor completely:

$8(x - y) + a(y - x)$

A. $(8 + a)(y - x)$

B. $(8 - a)(y - x)$

C. $(a - 8)(y - x)$

D. $(a - 8)(y + x)$

(Average) (Skill 1.14)

115. Which of the following is a factor of
$k^3 - m^3$?

A. $k^2 + m^2$

B. $k + m$

C. $k^2 - m^2$

D. $k - m$

(Rigorous) (Skill 1.19)

116. What is the solution set for the following equations?

$3x + 2y = 12$
$12x + 8y = 15$

A. All real numbers

B. $x = 4, y = 4$

C. $x = 2, y = -1$

D. $\varnothing$

(Rigorous) (Skill 1.19)

117. Solve for x and y:

$x = 3y + 7$
$7x + 5y = 23$

A. $(-1, 4)$

B. $(4, -1)$

C. $(\frac{-29}{7}, \frac{-26}{7})$

D. $(10, 1)$

(Rigorous) (Skill 1.19)

118. Solve the system of equations for *x, y,* and *z.*

$$3x + 2y - z = 0$$
$$2x + 5y = 8z$$
$$x + 3y + 2z = 7$$

A. (-1, 2, 1)

B. (1, 2, -1)

C. (-3, 4, -1)

D. (0, 1, 2)

(Rigorous) (Skill 1.20)

119. Find the zeroes of $f(x) = x^3 + x^2 - 14x - 24.$

A. 4, 3, 2

B. 3, -8

C. 7, -2, -1

D. 4, -3, -2

(Rigorous) (Skill 1.20)

120. The discriminant of a quadratic equation is evaluated and determined to be -3. The equation has:

A. One real root

B. One complex root

C. Two roots, both real

D. Two roots, both complex

(Rigorous) (Skill 3.3)

121. Which equation is graphed below?

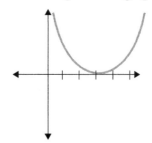

A. $y = 4 (x + 3)^2$

B. $y = 4 (x - 3)^2$

C. $y = 3 (x - 4)^2$

D. $y = 3 (x + 4)^2$

(Easy) (Skill 3.2)

122. Which set illustrates a function?

A. {(0,1) (0,2) (0,3) (0,4)}

B. {(3, 9) (-3, 9) (4, 16) (-4, 16)}

C. {(1, 2) (2, 3) (3, 4) (1, 4)}

D. {(2, 4) (3, 6) (4, 8) (4, 16)}

(Rigorous) (Skill 3.7)

123. Give the domain for the function over the set of real numbers:

$$y = \frac{3x + 2}{2x^2 - 3}$$

A. All real numbers

B. All real numbers, $x \neq 0$

C. All real numbers, $x \neq -2$ or 3

D. All real numbers, $x \neq \frac{\pm\sqrt{6}}{2}$

(Rigorous) (Skill 3.8)

124. If y varies inversely as x and x is 4 when y is 6, what is the constant of variation?

A. 2

B. 12

C. $\frac{3}{2}$

D. 24

(Rigorous) (Skill 3.8)

125. If y varies directly as x and x is 2 when y is 6, what is x when y is 18?

A. 3

B. 6

C. 26

D. 36

(Rigorous) (Skill 3.7)

126. State the domain of the function
$f(x) = \frac{3x - 6}{x^2 - 25}$

A. $x \neq 2$

B. $x \neq 5, -5$

C. $x \neq 2, -2$

D. $x \neq 5$

(Average) (Skill 3.5)

127. The volume of water flowing through a pipe varies directly with the square of the radius of the pipe. If the water flows at a rate of 80 liters per minute through a pipe with a radius of 4 cm, at what rate would water flow through a pipe with a radius of 3 cm?

A. 45 liters per minute

B. 6.67 liters per minute

C. 60 liters per minute

D. 4.5 liters per minute

(Average) (Skill 3.8)

128. Three less than four times a number is five times the sum of that number and 6. Which equation could be used to solve this problem?

A. $3 - 4n = 5(n + 6)$

B. $3 - 4n + 5n = 6$

C. $4n - 3 = 5n + 6$

D. $4n - 3 = 5(n + 6)$

(Average) (Skill 4A.7)

129. Find the median of the following set of data: 14   3   7   6   11   20

A. 9

B. 8.5

C. 7

D. 11

(Average) (Skill 4A.7)

130. Compute the median for the following data set:

{12, 19, 13, 16, 17, 14}

A. 14.5

B. 15.17

C. 15

D. 16

(Average) (Skill 4A.7)

131. Corporate salaries are listed for several employees. Which would be the best measure of central tendency?

$24,000	$24,000	$26,000
$28,000	$30,000	$120,000

A. Mean

B. Median

C. Mode

D. No difference

(Average) (Skill 4A.7)

132. Half the students in a class scored 80% on an exam, most of the rest scored 85% except for one student who scored 10%. Which would be the best measure of central tendency for the test scores?

A. Mean

B. Median

C. Mode

D. Either the median or the mode because they are equal

(Average) (Skill 4A.8)

133. A student scored in the 87th percentile on a standardized test. Which would be the best interpretation of his score?

A. Only 13% of the students who took the test scored higher

B. This student should be getting mostly Bs on his report card

C. This student performed below average on the test

D. This is the equivalent of missing 13 questions on a 100 question exam

(Easy) (Skill 4A.2)

134. Which statement is true about George's budget?

A. George spends the greatest portion of his income on food

B. George spends twice as much on utilities as he does on his mortgage

C. George spends twice as much on utilities as he does on food

D. George spends the same amount on food and utilities as he does on mortgage

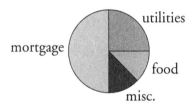

(Average) (Skill 4A.2)

135. What conclusion can be drawn from the graph below?

MLK Elementary Student Enrollment

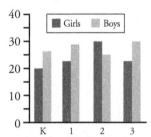

A. The number of students in first grade exceeds the number in second grade.

B. There are more boys than girls in the entire school

C. There are more girls than boys in the first grade

D. Third grade has the largest number of students

(Average) (Skill 4A.3)

136. Given a drawer with 5 black socks, 3 blue socks, and 2 red socks, what is the probability that you will draw two black socks in two draws in a dark room?

   A. $\frac{2}{9}$

   B. $\frac{1}{4}$

   C. $\frac{17}{18}$

   D. $\frac{1}{18}$

(Average) (Skill 4A.3)

137. A sack of candy has 3 peppermints, 2 butterscotch drops, and 3 cinnamon drops. One candy is drawn and replaced, then another candy is drawn; what is the probability that both will be butterscotch?

   A. $\frac{1}{2}$

   B. $\frac{1}{28}$

   C. $\frac{1}{4}$

   D. $\frac{1}{16}$

(Easy) (Skill 4A.3)

138. Given a spinner with the numbers one through eight, what is the probability that you will spin an even number or a number greater than four?

   A. $\frac{1}{4}$

   B. $\frac{1}{2}$

   C. $\frac{3}{4}$

   D. 1

(Rigorous) (Skill 4A.3)

139. If a horse will probably win three races out of ten, what are the odds that he will win?

   A. 3:10

   B. 7:10

   C. 3:7

   D. 7:3

(Rigorous) (Skill 4A.3)

140. How many ways are there to choose a potato and two green vegetables from a choice of three potatoes and seven green vegetables?

   A. 126

   B. 63

   C. 21

   D. 252

(Average) (Skill 4B.2)

141. Determine the number of subsets of set K. $K = \{4, 5, 6, 7\}$

   A. 15

   B. 16

   C. 17

   D. 18

(Average) (Skill 4B.4)

142. $\{1,4,7,10, \ldots\}$

   What is the 40th term in this sequence?

   A. 43

   B. 121

   C. 118

   D. 120

(Average) (Skill 4B.4)

143. {6, 11, 16, 21, . .}

Find the sum of the first 20 terms in the sequence.

A. 1070

B. 1176

C. 969

D. 1069

(Rigorous) (Skill 4B.4)

144. Find the sum of the first one hundred terms in the progression. (−6, −2, 2 . . .)

A. 19,200

B. 19,400

C. −604

D. 604

(Rigorous) (Skill 4B.7)

145. What would be the seventh term of the expanded binomial $(2a + b)^8$?

A. $2ab^7$

B. $41a^4b^4$

C. $112a^2b^6$

D. $16ab^7$

(Easy)(Domain V)

146. Which is the least appropriate strategy to emphasize when teaching problem solving?

A. Guess and check

B. Look for key words to indicate operations such as all together—add, more than, subtract, times, multiply

C. Make a diagram

D. Solve a simpler version of the problem

ANSWER KEY								
1. A	18. D	35. A	52. D	69. A	86. C	103. D	120. D	137. D
2. B	19. A	36. D	53. B	70. A	87. B	104. A	121. B	138. C
3. C	20. B	37. B	54. C	71. C	88. A	105. D	122. B	139. C
4. B	21. D	38. B	55. C	72. D	89. D	106. C	123. D	140. B
5. B	22. C	39. B	56. D	73. B	90. B	107. D	124. D	141. B
6. A	23. C	40. B	57. C	74. D	91. B	108. C	125. B	142. C
7. A	24. D	41. A	58. C	75. B	92. A	109. B	126. B	143. A
8. D	25. B	42. B	59. B	76. C	93. B	110. B	127. A	144. A
9. C	26. C	43. D	60. B	77. C	94. C	111. C	128. D	145. C
10. A	27. A	44. B	61. C	78. B	95. A	112. A	129. A	146. B
11. C	28. D	45. A	62. B	79. B	96. B	113. A	130. C	
12. D	29. C	46. C	63. A	80. D	97. B	114. C	131. B	
13. A	30. A	47. C	64. B	81. C	98. C	115. D	132. B	
14. D	31. A	48. B	65. B	82. A	99. C	116. D	133. A	
15. C	32.A	49. C	66. B	83. C	100. C	117. B	134. C	
16. B	33. D	50. C	67. C	84. D	101. D	118. A	135. B	
17. C	34. C	51. A	68. A	85. D	102. B	119. D	136. A	

RIGOR TABLE	
Rigor level	**Questions**
Easy 20%	5, 6, 7, 9, 10, 11, 14, 19, 34, 37, 41, 42, 43, 45, 47, 48, 65, 69, 76, 77, 78, 81, 83, 85, 98, 109, 122, 134, 138, 146
Average 40%	1, 2, 3, 4, 8, 15, 16, 18, 20, 21, 24, 27, 28, 29, 30, 31, 32, 35, 38, 39, 40, 46, 50, 52, 56, 59, 60, 62, 63, 68, 70, 71, 72, 74, 80, 82, 84, 86, 99, 102, 106, 108, 113, 114, 115, 127, 128, 129, 130, 131, 132, 133, 135, 136, 137, 141, 142, 143
Rigorous 40%	12, 13, 17, 22, 23, 25, 26, 33, 36, 44, 49, 51, 53, 54, 55, 57, 58, 61, 64, 66, 67, 73, 75, 79, 87, 88, 89, 90, 91, 92, 93, 94, 95, 96, 97, 100, 101, 103, 104, 105, 107, 110, 111, 112, 116, 117, 118, 119, 120, 121, 123, 124, 125, 126, 139, 140, 144, 145

Sample Questions with Rationales

The following represent one way to solve the problems and obtain a correct answer.

There are many other mathematically correct ways of determining the correct answer.

(Average)(Skill 1.1)

1. What would be the total cost of a suit for $295.99 and a pair of shoes for $69.95 including 6.5% sales tax?

    A. $389.73
    B. $398.37
    C. $237.86
    D. $315.23

Answer: A. $389.73

Before the tax, the total comes to $365.94. Then .065(365.94) = 23.79. With the tax added on, the total bill is 365.94 + 23.79 = $389.73. (Quicker way: 1.065(365.94) = 389.73.)

(Average) (Skill 1.5)

2. A student had 60 days to appeal the results of an exam. If the results were received on March 23, what was the last day that the student could appeal?

    A. May 21
    B. May 22
    C. May 23
    D. May 24

Answer: B. May 22

Recall: 30 days in April and 31 in March. 8 days in March + 30 days in April + 22 days in May brings him to a total of 60 days on May 22.

(Average) (Skill 1.1)

3. Evaluate: $\frac{1}{3} - \frac{1}{2} + \frac{1}{6}$

    A. $\frac{5}{6}$
    B. $\frac{2}{3}$
    C. 0
    D. 1

Answer: C. 0

$$\frac{1}{3} - \frac{1}{2} + \frac{1}{6} = \frac{2}{6} - \frac{3}{6} + \frac{1}{6} = \frac{2 - 3 + 1}{6} = 0$$

(Average) (Skill 1.19)

4. Solve for x: $3(5 + 3x) - 8 = 88$

    A. 30
    B. 9
    C. 4.5
    D. 27

Answer: B. 9

$3(5 + 3x) - 8 = 88$; $15 + 9x - 8 = 88$; $7 + 9x = 88$; $9x = 81$; $x = 9$

(Easy) (Skill 1.5)

5. Sandra has $34.00, Carl has $42.00. How much more does Carl have than Sandra? Which would be the best method for finding the answer?

    A. Addition
    B. Subtraction
    C. Division
    D. Both A and B are equally correct

Answer: B. Subtraction

PEMDAS

(Easy) (Skill 1.14)

6. Express in symbols: "x is greater than seven and less than or equal to fifteen."

   A. $7 < x \le 15$
   B. $7 > x \ge 15$
   C. $15 \le x < 7$
   D. $7 < x = 15$

   Answer: A. $7 < x \le 15$

(Easy) (Skill 1.6)

7. Which of the following is correct?

   A. $2365 > 2340$
   B. $0.75 > 1.25$
   C. $\frac{3}{4} < \frac{1}{16}$
   D. $-5 < -6$ *−6 −5 −4*

   Answer: A. $2365 > 2340$

(Average) (Skill 1.11)

8. Given that n is a positive even integer, $5n + 4$ will always be divisible by:

   A. 4   *$5(2) + 4 = 14$*
   B. 5
   C. $5n$ *$5(4) + 4 = 24$*
   D. 2

   Answer: D. 2

   $5n$ is always even and even number added to an even number is always an even number, thus divisible by 2.

(Easy) (Skill 1.2)

9. A student turns in a paper with this type of error:

   $7 + 16 \div 8 \times 2 = 8$ *11*
   $8 - 3 \times 3 + 4 = -5$ *3*

   In order to remediate this error, a teacher should:

   A. Review and drill basic number facts
   B. Emphasize the importance of using parentheses in simplifying expressions
   C. Emphasize the importance of working from left to right when applying the order of operations
   D. Do nothing; these answers are correct

   Answer: C. Emphasize the importance of working from left to right when applying the order of operations

(Easy) (Skill 1.7)

10. Given W = whole numbers *0, 1, 2, 3*
    N = natural numbers *1, 2, 3*
    Z = integers *−1, 0, 1*
    R = rational numbers *fraction/decimal π = 3.14*
    I = irrational numbers *√5*

    which of the following is not true?

    A. $R \subset I$
    B. $W \subset Z$
    C. $Z \subset R$
    D. $N \subset W$ *⊂ = are included*

    Answer: A. $R \subset I$

    The rational numbers are not a subset of the irrational numbers. All of the other statements are true.

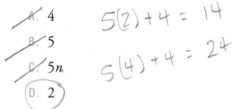

(Easy) (Skill 1.7)

11. **Which of the following is an irrational number?**

    A. .362626262...

    B. $4\frac{1}{3}$

    C. $\sqrt{5}$

    D. $-\sqrt{16}$

Answer: C. $\sqrt{5}$

$\sqrt{5}$ is an irrational number. A and B can both be expressed as fractions. D can be simplified to –4, an integer and rational number.

(Rigorous) (Skill 1.7)

12. **Which denotes a complex number?**

    A. 3.678678678...

    B. $-\sqrt{27}$

    C. $123^{\frac{1}{2}}$

    D. $(-100)^{\frac{1}{2}}$ $\sqrt{-100}$

Answer: D. $(-100)^{\frac{1}{2}}$

The square root of a negative number is a complex number. The complex number *i* is defined as the square root of –1. The exponent $\frac{1}{2}$ represents a square root.

(Rigorous) (Skill 1.7)

13. **Choose the correct statement:**

    A. Rational and irrational numbers are both proper subsets of the set of real numbers

    B. The set of whole numbers is a proper subset of the set of natural numbers

    C. The set of integers is a proper subset of the set of irrational numbers

    D. The set of real numbers is a proper subset of the natural, whole, integers, rational, and irrational numbers   *equal to*

Answer: A. Rational and irrational numbers are both proper subsets of the set of real numbers

A proper subset is completely contained in but not equal to the original set.

(Easy) (Skill 1.6)

14. **How many real numbers lie between –1 and +1?**

    A. 0

    B. 1

    C. 17

    D. An infinite number

Answer: D. An infinite number

There are an infinite number of real numbers between any two real numbers.

(Average) (Skill 1.6)

15. **Choose the set in which the members are not equivalent.**

    A. $\frac{1}{2}$, 0.5, 50%

    B. $\frac{10}{5}$, 2.0, 200%

    C. $\frac{3}{8}$, 0.385, 38.5%

    D. $\frac{7}{10}$, 0.7, 70%

    Answer: C. $\frac{3}{8}$, 0.385, 38.5%

    $\frac{3}{8}$ is equivalent to .375 and 37.5%.

(Average) (Skill 1.6)

16. **Change $.\overline{63}$ into a fraction in simplest form.**

    A. $\frac{63}{100}$

    B. $\frac{7}{11}$

    C. $6\frac{3}{10}$

    D. $\frac{2}{3}$

    Answer: B. $\frac{7}{11}$

    Let $N = .636363...$. Then multiplying both sides of the equation by 100 or 10^2 (because there are 2 repeated numbers), we get $100N = 63.636363...$ Then subtracting the two equations gives $99N = 63$ or $N = \frac{63}{99} = \frac{7}{11}$.

(Rigorous) (Skill 1.11)

17. **Which of the following is always composite if x is odd, y is even, and both x and y are greater than or equal to 2?**

    A. $x + y$

    B. $3x + 2y$

    C. $5xy$

    D. $5x + 3y$

Answer: C. $5xy$

Irrespective of the values of x and y and whether they are even or odd, $5xy$ will always be composite since it will always have the numbers 5, x, and y as factors.

(Average) (Skill 1.8)

18. **Express .0000456 in scientific notation.**

    A. 4.56×10^{-4}

    B. 45.6×10^{-6}

    C. 4.56×10^{-6}

    D. 4.56×10^{-5}

    Answer: D. 4.56×10^{-5}

    In scientific notation, the decimal point belongs to the right of the 4, the first significant digit. To get from 4.56×10^{-5} back to 0.0000456, we would move the decimal point 5 places to the left.

(Easy) (Skill 1.5)

19. **Mr. Brown feeds his cat premium cat food which costs $40 per month. Approximately how much will it cost to feed her for one year?**

    A. $500

    B. $400

    C. $80

    D. $4800

    Answer: A. $500

    $12(40) = 480$, which is closest to $500.

(Average) (Skill 1.8)

20. $(3.8 \times 10^{17}) \times (.5 \times 10^{-12})$

    A. 19×10^5

    B. 1.9×10^5

    C. 1.9×10^6

    D. 1.9×10^7

Answer: B. 1.9×10^5

Multiply the decimals and add the exponents.

(Average) (Skill 1.8)

21. 2^{-3} is equivalent to

    A. 0.8

    B. -0.8

    C. 125

    D. 0.125

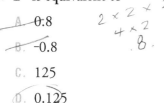

Answer: D. 0.125

Express as the fraction $\frac{1}{8}$, and then convert to a decimal.

(Rigorous) (Skill 1.8)

22. $\dfrac{3.5 \times 10^{-10}}{0.7 \times 10^{4}}$

    A. 0.5×10^6

    B. 5.0×10^{-6}

    C. 5.0×10^{-14}

    D. 0.5×10^{-14}

Answer: C. 5.0×10^{-14}

Divide the decimals and subtract the exponents.

(Rigorous) (Skill 1.17)

23. Simplify $\dfrac{\frac{3}{4}x^2y^{-3}}{\frac{2}{3}xy}$

    A. $\frac{1}{2}xy^{-4}$

    B. $\frac{1}{2}x^{-1}y^{-4}$

    C. $\frac{9}{8}xy^{-4}$

    D. $\frac{9}{8}xy^{-2}$

Answer: C. $\frac{9}{8}xy^{-4}$

Simplify the complex fraction by inverting the denominator and multiplying: $\frac{3}{4}\left(\frac{3}{2}\right) = \frac{9}{8}$, then subtract exponents to obtain the correct answer.

(Average) (Skill 1.17)

24. **Which of the following is incorrect?**

    A. $(x^2y^3)^2 = x^4y^6$

    B. $m^2(2n)^3 = 8m^2n^3$

    C. $\frac{(m^3n^4)}{(m^2n^2)} = mn^2$

    D. $(x + y^2)^2 = x^2 + y^4$

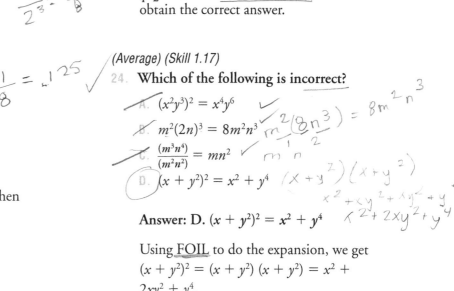

Answer: D. $(x + y^2)^2 = x^2 + y^4$

Using FOIL to do the expansion, we get $(x + y^2)^2 = (x + y^2)(x + y^2) = x^2 + 2xy^2 + y^4$.

(Rigorous) (Skill 1.8)

25. Evaluate: $3^{\frac{1}{2}} (9^{\frac{1}{3}})$

    A. $27^{\frac{5}{6}}$

    B. $9^{\frac{7}{12}}$

    C. $3^{\frac{5}{6}}$

    D. $3^{\frac{6}{7}}$

Answer: B. $9^{\frac{7}{12}}$

Getting the bases the same gives us $3^{\frac{1}{2}}3^{\frac{2}{3}}$. Adding exponents gives $3^{\frac{7}{6}}$. Then some additional manipulation of exponents produces $3^{\frac{7}{6}} = 3^{\frac{14}{12}} = (3^2)^{\frac{7}{12}} = 9^{\frac{7}{12}}$.

(Rigorous) (Skill 1.9)

26. Solve: $\sqrt{75} + \sqrt{147} - \sqrt{48}$

    A. 174

    B. $12\sqrt{3}$

    C. $8\sqrt{3}$

    D. 74

Answer: C. $8\sqrt{3}$

Simplify each radical by factoring out the perfect squares: $5\sqrt{3} + 7\sqrt{3} - 4\sqrt{3} = 8\sqrt{3}$

(Average) (Skill 1.9)

27. Simplify: $\sqrt{27} + \sqrt{75}$

    A. $8\sqrt{3}$

    B. 34

    C. $34\sqrt{3}$

    D. $15\sqrt{3}$

Answer: A. $8\sqrt{3}$

Simplifying radicals gives $\sqrt{27} + \sqrt{75} = 3\sqrt{3} + 5\sqrt{3} = 8\sqrt{3}$.

(Average) (Skill 1.11)

28. Given that x, y, and z are prime numbers, which of the following is true?

    A. $x + y$ is always prime

    B. xyz is always prime

    C. xy is sometimes prime

    D. $x + y$ is sometimes prime

Answer: D. $x + y$ is sometimes prime

$x + y$ is sometimes prime. B and C show the products of two numbers which are always composite. $x + y$ may be prime, but not always. Hence A is incorrect.

(Average) (Skill 1.11)

29. Find the GCF of $2^2 \times 3^2 \times 5$ and $2^2 \times 3 \times 7$.

    A. $2^5 \times 3^3 \times 5 \times 7$

    B. $2 \times 3 \times 5 \times 7$

    C. $2^2 \times 3$

    D. $2^3 \times 3^2 \times 5 \times 7$

Answer: C. $2^2 \times 3$

Choose the number of each prime factor that is in common.

(Average) (Skill 1.11)

30. Given even numbers x and y, which could be the LCM of x and y?

    A. $\frac{xy}{2}$

    B. $2xy$

    C. $4xy$

    D. xy

Answer: A. $\frac{xy}{2}$

Although choices B, C and D are common multiples, when both numbers are even, the product can be divided by two to obtain the least common multiple.

(Average) (Skill 1.2)

31. $24 - 3 \times 7 + 2 =$

A. 5
B. 149
C. -3
D. 189

Answer: A. 5

According to the order of operations, multiplication is performed first, and then addition and subtraction from left to right.

(Average) (Skill 1.2)

32. $7t - (4 \times 2t) + (3t \times 4) \div 2 =$

A. 5t
B. 0
C. 31t
D. 18t

Answer: A. 5t

First perform multiplication and division from left to right; $7t - 8t + 6t$, then add and subtract from left to right.

(Rigorous) (Skill 1.12)

33. Joe reads 20 words/min., and Jan reads 80 words/min. How many minutes will it take Joe to read the same number of words that it takes Jan 40 minutes to read?

A. 10
B. 20
C. 80
D. 160

Answer: D. 160

If Jan reads 80 words/minute, she will read 3200 words in 40 minutes.

$\frac{3200}{20} = 160$

At 20 words per minute, it will take Joe 160 minutes to read 3200 words.

(Easy) (Skill 1.12)

34. If three cups of concentrate are needed to make 2 gallons of fruit punch, how many cups are needed to make 5 gallons?

A. 6 cups
B. 7 cups
C. 7.5 cups
D. 10 cups

Answer: C. 7.5 cups

Set up the proportion $\frac{3}{2} = \frac{x}{5}$, cross multiply to obtain $15 = 2x$, then divide both sides by 2.

(Average) (Skill 1.12)

35. **A sofa sells for $520. If the retailer makes a 30% profit, what was the wholesale price?**

    A. $400

    B. $676

    C. $490

    D. $364

Answer: A. $400

Let x be the wholesale price, then $x + .30x = 520$, $1.30x = 520$. Divide both sides by 1.30.

(Rigorous) (Skill 1.16)

36. **Simplify:** $\frac{10}{1 + 3i}$

    A. $-1.25(1 - 3i)$

    B. $1.25(1 + 3i)$

    C. $1 + 3i$

    D. $1 - 3i$

Answer: D. $1 - 3i$

Multiplying numerator and denominator by the conjugate gives $\frac{10}{1 + 3i} \times \frac{1 - 3i}{1 - 3i} =$
$\frac{10(1 - 3i)}{1 - 9i^2} = \frac{10(1 - 3i)}{1 - 9(-1)} = \frac{10(1 - 3i)}{10} =$
$1 - 3i$.

(Easy) (Skill 1.3)

37. **Which statement is an example of the identity axiom of addition?**

    A. $3 + -3 = 0$

    B. $3x = 3x + 0$

    C. $3 \times \frac{1}{3} = 1$

    D. $3 + 2x = 2x + 3$

Answer: B. $3x = 3x + 0$

Illustrates the identity axiom of addition. A illustrates additive inverse, C illustrates the multiplicative inverse, and D illustrates the commutative axiom of addition.

(Average) (Skill 1.3)

38. **Which axiom is incorrectly applied?**

    $3x + 4 = 7$

    Step a. $3x + 4 - 4 = 7 - 4$
    additive equality

    Step b. $3x + 4 - 4 = 3$
    commutative axiom of addition

    Step c. $3x + 0 = 3$
    additive inverse

    Step d. $3x = 3$
    additive identity

    A. Step a

    B. Step b

    C. Step c

    D. Step d

Answer: B. Step b

In simplifying from step a to step b, 3 replaced $7 - 4$, therefore the correct justification would be subtraction or substitution.

(Average) (Skill 1.3)

39. **Which of the following sets is closed under division?**

    A. Integers

    B. Rational numbers

    C. Natural numbers

    D. Whole numbers

Answer: B. Rational numbers

In order to be closed under division, when any two members of the set are divided, the answer must be contained in the set. This is not true for integers, natural, or whole numbers as illustrated by the counter example $\frac{11}{2} = 5.5$.

(Average) (Skill 1.4)

40. Which of the following does not correctly relate an inverse operation?

   A. $a - b = a + {}^{-}b$

   B. $a \times b = b \div a$

   C. $\sqrt{a^2} = a$

   D. $a \times \frac{1}{a} = 1$

Answer: B. $a \times b = b \div a$

B is always false. A, C, and D illustrate various properties of inverse relations.

(Easy) (Skill 1.3)

41. Given a, b, y, and z are real numbers and $ay + b = z$, prove $y = \frac{z + (-b)}{a}$

STATEMENT	REASON
1. $ay + b = z$	Given
2. $-b$ is a real number	Closure
3. $(ay + b) + {}^-b = z + {}^-b$	Identity property of addition
4. $ay + (b + {}^-b) = z + {}^-b$	Associative
5. $ay + 0 = z + {}^-b$	Additive inverse
6. $ay = z + {}^-b$	Identity property of addition
7. $y = \frac{z + (-b)}{a}$	Division

Which reason is incorrect for the corresponding statement?

A. Step 3

B. Step 4

C. Step 5

D. Step 6

Answer: A. Step 3

(Easy) (Skill 1.3)

42. Which of the following sets is closed under division?

   I. $\{\frac{1}{2}, 1, 2, 4\}$

   II. $\{{}^-1, 1\}$

   III. $\{{}^-1, 0, 1\}$

   A. I only

   B. II only

   C. III only

   D. I and II

Answer: B. II only

I is not closed because $\frac{4}{.5} = 8$ and 8 is not in the set. III is not closed because $\frac{1}{0}$ is undefined. II is closed because $\frac{-1}{1} = {}^-1$, $\frac{1}{-1} = {}^-1$, $\frac{1}{1} = 1$, $\frac{-1}{-1} = 1$ and all the answers are in the set.

(Easy) (Skill 1.4)

43. Which of the following illustrates an inverse property?

   A. $a + b = a - b$

   B. $a + b = b + a$ *com.*

   C. $a + 0 = a$ *id.*

   D. $a + (-a) = 0$ *inverse*

Answer: D. $a + (-a) = 0$

Because $a + (-a) = 0$ is a statement of the Additive Inverse Property of Algebra.

(Rigorous) (Skill 1.19)

44. A boat travels 30 miles upstream in three hours. It makes the return trip in one and a half hours. What is the speed of the boat in still water?

   A. 10 mph

   B. 15 mph

   C. 20 mph

   D. 30 mph

Answer: B. 15 mph

Let x = the speed of the boat in still water and c = the speed of the current.

	Rate	Time	Distance
Upstream	$x - c$	3	30
Downstream	$x + c$	1.5	30

Solve the system:
$3x - 3c = 30$
$1.5x + 1.5c = 30$

(Easy) (Skill 2.5)

45. Given $XY \cong YZ$ and $\angle AYX \cong \angle AYZ$. Prove $\triangle AYZ \cong \triangle AYX$.

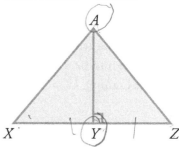

1. $XY \cong YZ$

2. $\angle AYX \cong \angle AYZ$

3. $AY \cong AY$

4. $\triangle AYZ \cong \triangle AYX$

Which property justifies step 3?

   A. Reflexive

   B. Symmetric

   C. Transitive

   D. Identity

Answer: A. Reflexive

The reflexive property states that every number or variable is equal to itself and every segment is congruent to itself.

(Average) (Skill 2.6)

46. Given $l_1 \parallel l_2$ prove $\angle b \cong \angle e$.

1. $\angle b \cong \angle d$ vertical angle theorem

2. $\angle d \cong \angle e$ alternate interior angle theorem

3. $\angle b \cong \angle e$ symmetric axiom of equality

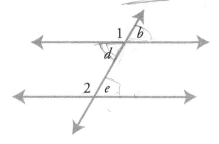

Which step is incorrectly justified?

A. Step 1

B. Step 2

C. Step 3

D. No error

Answer: C. Step 3

Step 3 can be justified by the transitive property.

(Easy) (Skill 2.8)

47. Which of the following shapes is not a parallelogram?

I

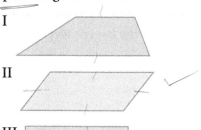

II

III

A. I & III

B. II & III

C. I

D. I, II & III

Answer: C. I

A parallelogram is a quadrilateral with two pairs of parallel sides.

(Easy) (Skill 2.1)

48. 4 lbs 4 oz is equal to:

A. 20 oz

B. 4.25 lb

C. 4.33 oz

D. 64 oz

Answer: B. 4.25 lb

Since 16 oz = 1 lb, 4 oz = 0.25 lb

(Rigorous) (Skill 2.5)

49. Given similar polygons with corresponding sides of lengths 9 and 15, find the perimeter of the smaller polygon if the perimeter of the larger polygon is 150 units.

  A. 54

  B. 135

  C. 90

  D. 126

Answer: C. 90

The perimeters of similar polygons are directly proportional to the lengths of their sides, therefore $\frac{9}{15} = \frac{x}{150}$. Cross multiply to obtain $1350 = 15x$, then divide by 15 to obtain the perimeter of the smaller polygon.

(Average) (Skill 2.6)

50. Given $l_1 \parallel l_2$ which of the following is true?

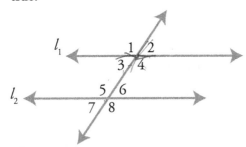

  A. $\angle 1$ and $\angle 8$ are congruent and alternate interior angles

  B. $\angle 2$ and $\angle 3$ are congruent and corresponding angles

  C. $\angle 3$ and $\angle 4$ are adjacent and supplementary angles

  D. $\angle 3$ and $\angle 5$ are adjacent and supplementary angles

Answer: C. $\angle 3$ and $\angle 4$ are adjacent and supplementary angles

The angles in A are exterior. In B, the angles are vertical. The angles in D are consecutive, not adjacent.

(Rigorous) (Skill 2.9)

51. Given the regular hexagon below, determine the measure of angle 1.

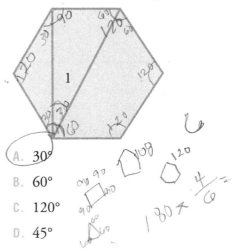

  A. 30°

  B. 60°

  C. 120°

  D. 45°

Answer: A. 30°

Each interior angle of the hexagon measures 120°. The isosceles triangle on the left has angles which measure 120, 30, and 30. By alternate interior angle theorem, $\angle 1$ is also 30.

(Average) (Skill 2.5)

52. Given $QS \cong TS$ and $RS \cong US$, prove $\triangle QRS \cong \triangle TUS$.

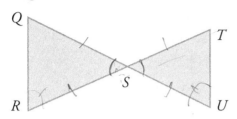

1. $QS \cong TS$	1. Given
2. $RS \cong US$	2. Given
3. $\angle TSU \cong \angle QSR$	3. ?
4. $\triangle TSU \cong \triangle QSR$	4. SAS

Give the reason which justifies step 3.

A. Congruent parts of congruent triangles are congruent

B. Reflexive axiom of equality

C. Alternate interior angle theorem

D. Vertical angle theorem

Answer: D. Vertical angle theorem

Angles formed by intersecting lines are called vertical angles and are congruent.

(Rigorous) (Skill 2.7)

53. In the figure below, what is the value of x?

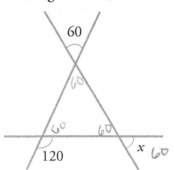

Note: Figure not drawn to scale.

A. 50

B. 60

C. 75

D. 80

Answer: B. 60

The angles within the triangle make up 180°. Opposite angles are equal, therefore, the angle opposite the 60° angle is also 60°. Adjacent angles add to 180° (straight line). Therefore, the angle inside the triangle adjacent to the 120° angle is 60°. The third angle in the triangle would then be 60° (180 − 60 − 60). Since x is opposite this third angle, it would also be 60°.

(Rigorous) (Skill 2.12)

54. **Line p has a negative slope and passes through the point $(0, 0)$. If line q is perpendicular to line p, which of the following must be true?**

    A. Line q has a negative y-intercept.

    B. Line q passes through the point $(0, 0)$

    C. Line q has a positive slope.

    D. Line q has a positive y-intercept.

Answer: C. Line q has a positive slope.

Draw a picture to help you visualize the problem.

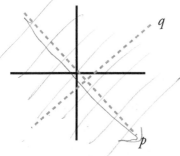

Choices (A) and (D) are not correct because line q could have a positive or a negative y-intercept. Choice (B) is incorrect because line q does not necessarily pass through $(0, 0)$. Since line q is perpendicular to line p. which has a negative slope, it must have a positive slope.

(Rigorous) (Skill 2.12)

55. **What is the slope of any line parallel to the line $2x + 4y = 4$?**

    A. -2

    B. -1

    C. $-\frac{1}{2}$

    D. 2

$4y = -2x + 4$

$y = -\frac{1}{2}x + 1$

Answer: C. $-\frac{1}{2}$

The formula for slope is $y = mx + b$, where m is the slope. Lines that are parallel have the same slope.

$2x + 4y = 4$

$4y = -2x + 4$

$y = \frac{-2x}{4} + 1$

$y = \frac{-1}{2}x + 1$

(Average) (Skill 2.5)

56. **Prove $\triangle HYM \cong \triangle KZL$, given $XZ \cong XY$, $\angle L \cong \angle M$ and $YL \cong MZ$**

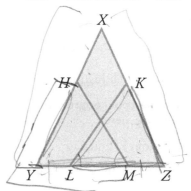

1. $XZ \cong XY$	1. Given
2. $\angle Y \cong \angle Z$	2. ?
3. $\angle L \cong \angle M$	3. Given
4. $YL \cong MZ$	4. Given
5. $LM \cong LM$	5. ?
6. $YM \cong LZ$	6. Add
7. $\triangle HYM \cong \triangle KZL$	7. ASA

Which could be used to justify steps 3 and 5?

A. CPCTC, Identity

B. Isosceles Triangle Theorem, Identity

C. SAS, Reflexive

D. Isosceles Triangle Theorem, Reflexive

Answer: D. Isosceles Triangle Theorem, Reflexive

The isosceles triangle theorem states that the base angles are congruent, and the reflexive property states that every segment is congruent to itself.

(Rigorous) (Skill 2.9)

57. **What is the degree measure of an interior angle of a regular 10-sided polygon?**

    A. 18°

    B. 36°

    C. 144°

    D. 54°

    Answer: C. 144°

    Formula for finding the measure of each interior angle of a regular polygon with n sides is $\frac{(n-2)180}{n}$. For $n = 10$, we get $\frac{8(180)}{10} = 144$.

(Rigorous) (Skill 2.7)

58. **Which of the following statements is true about the number of degrees in each angle?**

    A. $a + b + c = 180°$

    B. $a = e$

    C. $b + c = e$

    D. $c + d = e$

    Answer: C. $b + c = e$

    In any triangle, an exterior angle is equal to the sum of the remote interior angles.

(Average) (Skill 2.5)

59. **What method could be used to prove the below triangles congruent?**

    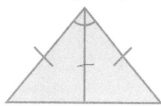

    A. SSS

    B. SAS

    C. AAS

    D. SSA

    Answer: B. SAS

    Use SAS with the last side being the vertical line common to both triangles.

(Average) (Skill 2.5)

60. **Which postulate could be used to prove $\triangle ABD \cong \triangle CEF$, given $BC \cong DE$, $\angle C \cong \angle D$, and $AD \cong CF$?**

    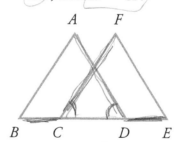

    A. ASA

    B. SAS

    C. SAA

    D. SSS

    Answer: B. SAS

    To obtain the final side, add CD to both BC and ED.

(Rigorous) (Skill 2.5)

61. **Which theorem can be used to prove**
    $\triangle BAK \cong \triangle MKA$?

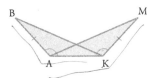

   A. SSS

   B. ASA

   C. SAS

   D. AAS

   Answer: C. SAS

   Since side AK is common to both triangles, the triangles can be proved congruent by using the Side-Angle-Side Postulate.

(Average) (Skill 2.5)

62. **Given similar polygons with corresponding sides 6 and 8, what is the area of the smaller if the area of the larger is 64?**

   A. 48

   B. 36

   C. 144

   D. 78

   Answer: B. 36

   In similar polygons, the areas are proportional to the squares of the sides. $\frac{36}{64} = \frac{x}{64}$

(Average) (Skill 2.5)

63. **In similar polygons, if the perimeters are in a ratio of $x : y$, the sides are in a ratio of:**

   A. $x : y$

   B. $x^2 : y^2$

   C. $2x : y$

   D. $\frac{1}{2}x : y$

   Answer: A. x : y

   The sides are in the same ratio.

(Rigorous) (Skill 2.7)

64. **ALM is a right triangle with the right angle at A. Given altitude AK with measurements as indicated, determine the length of AK.**

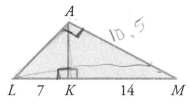

   A. 98

   B. $7\sqrt{2}$

   C. $\sqrt{21}$

   D. $7\sqrt{3}$

   Answer: B. $7\sqrt{2}$

   In a right triangle, the length of the altitude to the hypotenuse is the geometric mean between each segment of the hypotenuse. So AK2 = 7 × 14.

(Easy) (Skill 2.2)

65. The following diagram is most likely used in deriving a formula for which of the following?

   A. The area of a rectangle

   B. The area of a triangle

   C. The perimeter of a triangle

   D. The surface area of a prism

Answer: B. The area of a triangle

(Rigorous) (Skill 2.2)

66. Compute the area of the shaded region, given a radius of 5 meters. O is the center.

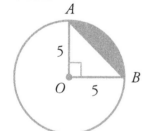

   A. 7.13 cm²

   B. 7.13 m²

   C. 78.5 m²

   D. 19.63 m²

Answer: B. 7.13 m²

Area of triangle AOB is .5(5)(5) = 12.5 square meters. Since $\frac{90}{360}$ = .25, the area of sector AOB (pie-shaped piece) is approximately .25(π)5² = 19.63. Subtracting the triangle area from the sector area to get the area of segment AB, we get approximately 19.63 − 12.5 = 7.13 square meters.

(Rigorous) (Skill 2.3)

67. Ginny and Nick head back to their respective colleges after being home for the weekend. They leave their house at the same time and drive for 4 hours. Ginny drives due south at the average rate of 60 miles per hour and Nick drives due east at the average rate of 60 miles per hour. What is the straight-line distance between them, in miles, at the end of the 4 hours?

   A. $120\sqrt{2}$

   B. 240

   C. $240\sqrt{2}$

   D. 288

Answer: C. $240\sqrt{2}$

Draw a picture.

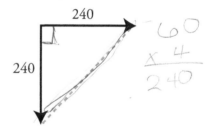

We have a right triangle, so we can use the Pythagorean Theorem to find the distance between the two points.

$240^2 + 240^2 = c^2$
$(2)240^2 = c^2$
$240\sqrt{2} = c$

(Average) (Skill 2.3)

68. If $AC = 12$, determine BC.

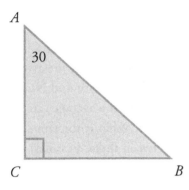

A. 6

B. 4

C. $6\sqrt{3}$

D. $3\sqrt{6}$

Answer: A. 6

In a 30-60-90 right triangle, the leg opposite the 30° angle is half the length of the hypotenuse.

(Easy) (Skill 2.8)

69. Given that $QO \perp NP$ and $QO = NP$, quadrilateral $NOPQ$ can most accurately be described as a

A. Parallelogram

B. Rectangle

C. Square

D. Rhombus

Answer: A. Parallelogram

In an ordinary parallelogram, the diagonals are not perpendicular or equal in length. In a rectangle, the diagonals are not necessarily perpendicular. In a rhombus, the diagonals are not equal in length. In a square, the diagonals are both perpendicular and congruent.

(Average) (Skill 2.7)

70. **Choose the correct statement concerning the median and altitude in a triangle.**

A. The median and altitude of a triangle may be the same segment

B. The median and altitude of a triangle are always different segments

C. The median and altitude of a right triangle is always the same segment

D. The median and altitude of an isosceles triangle is always the same segment

Answer: A. The median and altitude of a triangle may be the same segment

The most one can say with certainty is that the median (segment drawn to the midpoint of the opposite side) and the altitude (segment drawn perpendicular to the opposite side) of a triangle may coincide, but they more often do not. In an isosceles triangle, the median and the altitude to the base are the same segment.

(Average) (Skill 2.12)

71. **Find the distance between (3, 7) and (-3, 4).**

A. 9

B. 45

C. $3\sqrt{5}$

D. $5\sqrt{3}$

Answer: C. $3\sqrt{5}$

Using the distance formula

$$\sqrt{[3 - (-3)]^2 + (7 - 4)^2}$$
$$= \sqrt{36 + 9}$$
$$= 3\sqrt{5}$$

(Average) (Skill 2.12)

72. **Find the midpoint of (2, 5) and (7, –4).**

   A. (9, –1)

   B. (5, 9)

   C. $\left(\frac{9}{2}, -\frac{1}{2}\right)$

   D. $\left(\frac{9}{2}, \frac{1}{2}\right)$

   Answer: D. $\left(\frac{9}{2}, \frac{1}{2}\right)$

   Using the midpoint formula $x = \frac{(2 + 7)}{2}$
   $y = \frac{(5 + -4)}{2}$

(Rigorous) (Skill 2.12)

73. **Given segment *AC* with *B* as its midpoint find the coordinates of *C* if *A* = (5, 7) and *B* = (3, 6.5).**

   A. (4, 6.5)

   B. (1, 6)

   C. (2, 0.5)

   D. (16, 1)

   Answer: B. (1, 6)

(Average) (Skill 2.10)

74. **What is the measure of major arc *AL*?**

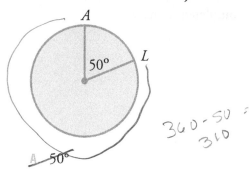

   A. 50°

   B. 25°

   C. 100°

   D. 310°

   Answer: D. 310°

   Minor arc *AL* measures 50°, the same as the central angle. To determine the measure of the major arc, subtract from 360.

(Rigorous) (Skill 2.10)

75. **If arc *KR* = 70° what is the measure of ∠*M*?**

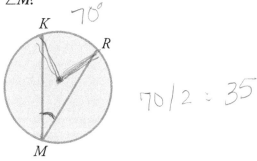

   A. 290°

   B. 35°

   C. 140°

   D. 110°

   Answer: B. 35°

   An inscribed angle is equal to one half the measure of the intercepted arc.

(Easy) (Skill 2.6)

76. **The following construction can be completed to make**

A. An angle bisector

B. Parallel lines

C. A perpendicular bisector

D. Skew lines

Answer: C. A perpendicular bisector

The points marked *C* and *D* are the intersection of the circles with centers *A* and *B*.

(Easy) (Skill 2.7)

77. **A line from *R* to *K* will form**

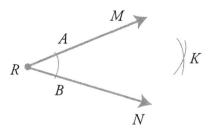

A. An altitude of *RMN*

B. A perpendicular bisector of *MN*

C. A bisector of *MRN*

D. A vertical angle

Answer: C. A bisector of *MRN*

Using a compass, point *K* is found to be equidistant from *A* and *B*.

(Easy) (Skill 2.4)

78. **An isosceles triangle has:**

A. Three equal sides

B. Two equal sides

C. No equal sides

D. Two equal sides in some cases, no equal sides in others

Answer: B. Two equal sides

(Rigorous) (Skill 2.10)

79. **What is the measure of minor arc *AD*, given measure of arc *PS* is 40° and $m < K = 10$?**

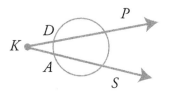

A. 50°

B. 20°

C. 30°

D. 25°

Answer: B. 20°

The formula relating the measure of angle *K* and the two arcs it intercepts is $m\angle K = \frac{1}{2}(mPS - mAD)$. Substituting the known values, we get $10 = \frac{1}{2}(40 - mAD)$. Solving for *mAD* gives an answer of 20 degrees.

(Average) (Skill 2.6)

80. Choose the diagram which illustrates the construction of a perpendicular to the line at a given point on the line.

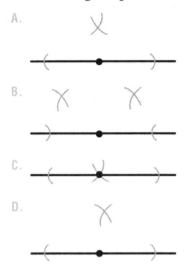

A.

B.

C.

D.

Answer: D.

Given a point on a line, place the compass point there and draw two arcs intersecting the line in two points, one on either side of the given point. Then using any radius larger than half the new segment produced, and with the pointer at each end of the new segment, draw arcs which intersect above the line. Connect this new point with the given point.

(Easy) (Skill 2.1)

81. Seventh grade students are working on a project using nonstandard measurement. Which would not be an appropriate instrument for measuring the length of the classroom?

A. A student's foot

B. A student's arm span

C. A student's jump

D. All are appropriate

Answer: C. A student's jump

While a student's foot or student's arm span has a fixed length, a student's jump can vary in length and would therefore not be an appropriate unit.

(Average) (Skill 2.1)

82. The speed of light in space is about 3×10^8 meters per second. Express this in Kilometers per hour.

A. 1.08×10^9 Km/hr

B. 3.0×10^{11} Km/hr

C. 1.08×10^{12} Km/hr

D. 1.08×10^{15} Km/hr

Answer: A. 1.08×10^9 KM/hr

$$3 \times 10^8 \tfrac{m}{s} = 3 \times 10^8 \tfrac{m}{s} \times \frac{1 Km}{1000 m} \times \frac{3600 s}{1 \, hr}$$
$$= 108 \times 10^7 \tfrac{Km}{hr} = 1.08 \times 10^9 \tfrac{Km}{hr}$$

(Easy) (Skill 2.1)

83. **The mass of a Chips Ahoy cookie would be close to:**

    A. 1 kilogram

    B. 1 gram

    C. 15 grams

    D. 15 milligrams

    Answer: C. 15 grams

    Since an ordinary cookie would not weigh as much as 1 kilogram, or as little as 1 gram or 15 milligrams, the only reasonable answer is 15 grams.

(Average) (Skill 2.13)

84. **A man's waist measures 90 cm. What is the greatest possible error for the measurement?**

    A. ±1 m

    B. ±8 cm

    C. ±1 cm

    D. ±5 mm

    Answer: D. ±5 mm

(Easy) (Skill 2.1)

85. **3 km is equivalent to:**

    A. 300 cm

    B. 300 m

    C. 3000 cm

    D. 3000 m

    Answer: D. 3000 m

    To change kilometers to meters, move the decimal 3 places to the right.

(Average) (Skill 2.1)

86. **4 square yards is equivalent to:**

    A. 12 square feet

    B. 48 square feet

    C. 36 square feet

    D. 108 square feet

    Answer: C. 36 square feet

    There are 9 square feet in a square yard.

(Rigorous) (Skill 2.2)

87. **If a circle has an area of 25 cm², what is its circumference to the nearest tenth of a centimeter?**

    A. 78.5 cm

    B. 17.7 cm

    C. 8.9 cm

    D. 15.7 cm

    Answer: B. 17.7 cm

    Find the radius by solving $\pi r^2 = 25$. Then substitute $r = 2.82$ into $C = 2\pi r$ to obtain the circumference.

(Rigorous) (Skill 2.2)

88. **Find the area of the figure below.**

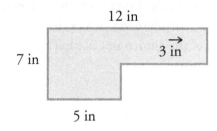

A. 56 in²

B. 27 in²

C. 71 in²

D. 170 in²

Answer: A. 56 in²

Divide the figure into two rectangles with a horizontal line. The area of the top rectangle is 36 in², and the bottom is 20 in².

(Rigorous) (Skill 2.2)

89. **Find the area of the shaded region given square *ABCD* with side *AB* = 10m and circle *E*.**

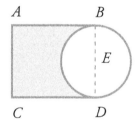

A. 178.5 m²

B. 139.25 m²

C. 71 m²

D. 60.75 m²

Answer: D. 60.75 m²

Find the area of the square $10^2 = 100$, then subtract $\frac{1}{2}$ the area of the circle. The

area of the circle is $\pi r^2 = (3.14)(5)(5)$ = 78.5. Therefore the area of the shaded region is $100 - 39.25 = 60.75$.

(Rigorous) (Skill 2.2)

90. **Compute the area of the polygon shown below.**

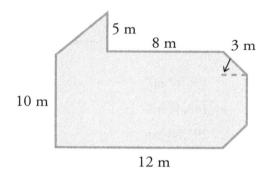

A. 178 m²

B. 154 m²

C. 43 m²

D. 188 m²

Answer: B. 154 m²

Divide the figure into a triangle, a rectangle and a trapezoid. The area of the triangle is $\frac{1}{2} bh = \frac{1}{2} (4)(5) = 10$. The area of the rectangle is $bh = 12(10) = 120$. The area of the trapezoid is $\frac{1}{2}(b + B) h = \frac{1}{2}(6 + 10)(3) = \frac{1}{2}(16)(3) = 24$. Thus, the area of the figure is $10 + 120 + 24 = 154$.

(Rigorous) (Skill 2.2)

91. **Find the area of the figure pictured below.**

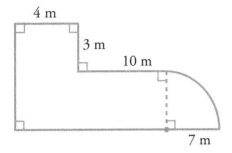

A. 136.47 m²

B. 148.48 m²

C. 293.86 m²

D. 178.47 m²

Answer: B. 148.48 m²

Divide the figure into 2 rectangles and one quarter circle. The tall rectangle on the left will have dimensions 10 by 4 and area 40. The rectangle in the center will have dimensions 7 by 10 and area 70. The quarter circle will have area $.25(\pi)7^2$ = 38.48. The total area is therefore approximately 148.48.

(Rigorous) (Skill 2.2)

92. **Given a 30 meter × 60 meter garden with a circular fountain with a 5 meter radius, calculate the area of the portion of the garden not occupied by the fountain.**

A. 1721 m²

B. 1879 m²

C. 2585 m²

D. 1015 m²

Answer: A. 1721 m²

Find the area of the garden and then subtract the area of the fountain: $30(60) - \pi(5)^2$ or approximately 1721 square meters.

(Rigorous) (Skill 2.2)

93. **Determine the area of the shaded region of the trapezoid in terms of x and y.**

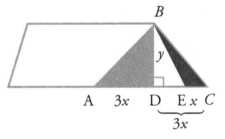

A. $4xy$

B. $2xy$

C. $3x^2y$

D. There is not enough information given

Answer: B. 2xy

To find the area of the shaded region, find the area of triangle *ABC* and then subtract the area of triangle *DBE*. The area of triangle *ABC* is $.5(6x)(y) = 3xy$. The area of triangle *DBE* is $.5(2x)(y) = xy$. The difference is $2xy$.

(Rigorous) (Skill 2.2)

94. **If the radius of a right circular cylinder is doubled, how does its volume change?**

    A. No change

    B. Also is doubled

    C. Four times the original

    D. Pi times the original

Answer: C. Four times the original

If the radius of a right circular cylinder is doubled, the volume is multiplied by four because in the formula, the radius is squared, therefore the new volume is 2 × 2 or four times the original.

(Rigorous) (Skill 2.2)

95. **Determine the volume of a sphere to the nearest cm if the surface area is 113 cm².**

    A. 113 cm³

    B. 339 cm³

    C. 37.7 cm³

    D. 226 cm³

Answer: A. 113 cm³

Solve for the radius of the sphere using $A = 4\pi r^2$. The radius is 3. Then, find the volume using $\frac{4}{3}\pi r^3$. Only when the radius is 3 are the volume and surface area equivalent.

(Rigorous) (Skill 2.2)

96. **Compute the surface area of the prism.**

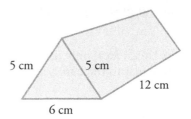

    A. 204

    B. 216

    C. 360

    D. 180

Answer: B. 216

There are five surfaces which make up the prism. The bottom rectangle has area 6 × 12 = 72. The sloping sides are two rectangles each with an area of 5 × 12 = 60. The height of the triangles is determined to be 4 using the Pythagorean theorem. Therefore each triangle has area $\frac{1}{2}bh = \frac{1}{2}(6)(4) = 12$. Thus, the surface area is 72 + 60 + 60 + 12 + 12 = 216.

(Rigorous) (Skill 2.2)

97. **If the base of a regular square pyramid is tripled, how does its volume change?**

    A. Double the original

    B. Triple the original

    C. Nine times the original

    D. No change

Answer: B. Triple the original

Using the general formula for a pyramid $V = \frac{1}{3}bh$, since the base is tripled and is not squared or cubed in the formula, the volume is also tripled.

(Easy) (Skill 2.2)

98. How does lateral area differ from total surface area in prisms, pyramids, and cones?

A. For the lateral area, only use surfaces perpendicular to the base

B. They are both the same

C. The lateral area does not include the base

D. The lateral area is always a factor of pi

Answer: C. The lateral area does not include the base

The lateral area does not include the base.

(Average) (Skill 2.2)

99. If the area of the base of a cone is tripled, the volume will be

A. The same as the original

B. 9 times the original

C. 3 times the original

D. 3 pi times the original

Answer: C. 3 times the original

The formula for the volume of a cone is $V = \frac{1}{3} Bh$, where B is the area of the circular base and h is the height. If the area of the base is tripled, the volume becomes $V = \frac{1}{3} (3B)h = Bh$, or three times the original volume.

(Rigorous) (Skill 2.2)

100. Find the height of a box with surface area of 94 sq. ft. with a width of 3 feet and a depth of 4 feet.

A. 3 ft.

B. 4 ft.

C. 5 ft

D. 6 ft.

Answer: C. 5 ft.

$$94 = 2(3h) + 2(4h) + 2(12)$$
$$94 = 6h + 8h + 24$$
$$94 = 14h + 24$$
$$70 = 14h$$
$$5 = h$$

(Rigorous) (Skill 3.1)

101. $f(x) = 3x - 2; f^{-1}(x) =$

A. $3x + 2$

B. $\frac{x}{6}$

C. $2x - 3$

D. $\frac{(x + 2)}{3}$

Answer: D. $\frac{(x + 2)}{3}$

To find the inverse, $f^{-1}(x)$, of the given function, reverse the variables in the given equation, $y = 3x - 2$, to get $x = 3y - 2$. Then solve for y as follows:

$x + 2 = 3y$, and $y = \frac{x + 2}{3}$

(Average) (Skill 1.19)

102. **Solve for x: $3x + 5 \geq 8 + 7x$**

A. $x \geq -\frac{3}{4}$

B. $x \leq -\frac{3}{4}$

C. $x \geq \frac{3}{4}$

D. $x \leq \frac{3}{4}$

Answer: B. $x \leq -\frac{3}{4}$

Using additive equality, $-3 \geq 4x$. Divide both sides by 4 to obtain $-\frac{3}{4} \geq x$. Carefully determine which answer choice is equivalent.

(Rigorous) (Skill 1.21)

103. **Solve for x: $\left| 2x + 3 \right| > 4$**

A. $\frac{7}{2} > x > \frac{1}{2}$

B. $\frac{1}{2} > x > \frac{7}{2}$

C. $x < \frac{7}{2}$ or $x > \frac{1}{2}$

D. $x < -\frac{7}{2}$ or $x > \frac{1}{2}$

Answer: D. $x < -\frac{7}{2}$ or $x > \frac{1}{2}$

The quantity within the absolute value symbols must be either > 4 or < -4. Solve the two inequalities $2x + 3 > 4$ or $2x + 3 < -4$

(Rigorous) (Skill 1.19)

104. **Graph the solution: $\left| x \right| + 7 < 13$**

A.

B.

C.

D.

Answer: A.

Solve by adding -7 to each side of the inequality. Since the absolute value of x is less than 6, x must be between -6 and 6. The end points are not included so the circles on the graph are hollow.

(Rigorous) (Skill 1.14)

105. **Solve for v_0 : $d = at(v_t - v_0)$**

A. $v_0 = atd - v_t$

B. $v_0 = d - atv_t$

C. $v_0 = atv_t - d$

D. $v_0 = \frac{(atv_t - d)}{at}$

Answer: D. $v_0 = \frac{(atv_t - d)}{at}$

Using the Distributive Property and other properties of equality to isolate v_0 gives $d = atv_t - atv_0$, $atv_0 = atv_t - d$, $v_0 = \frac{(atv_t - d)}{at}$.

(Average) (Skill 1.21)

106. **Solve for x: $18 = 4 + \left| 2x \right|$**

A. $\{-11, 7\}$

B. $\{-7, 0, 7\}$

C. $\{-7, 7\}$

D. $\{-11, 11\}$

Answer: C. $\{-7, 7\}$

Using the definition of absolute value, two equations are possible: $18 = 4 + 2x$ or $18 = 4 - 2x$. Solving for x gives $x = 7$ or $x = -7$.

(Rigorous) (Skill 3.3)

107. **Which graph represents the solution set for $x^2 - 5x > -6$?**

A.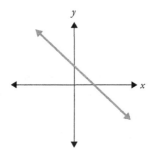
-2 0 2

B.
-3 0 3

C.
-2 0 2

D.
0 2 3

Answer: D.

Rewriting the inequality gives $x^2 - 5x + 6 > 0$. Factoring gives $(x - 2)(x - 3) > 0$. The two cut-off points on the number line are now at $x = 2$ and $x = 3$. Choosing a random number in each of the three parts of the number line, we test them to see if they produce a true statement. If $x = 0$ or $x = 4$, $(x - 2)(x - 3) > 0$ is true. If $x = 2.5$, $(x - 2)(x - 3) > 0$ is false. Therefore the solution set is all numbers smaller than 2 or greater than 3.

(Average) (Skill 3.3)

108. **Which equation is represented by the below graph?**

A. $x - y = 3$

B. $x - y = -3$

C. $x + y = 3$

D. $x + y = -3$

Answer: C. $x + y = 3$

By looking at the graph, we can determine the slope to be -1 and the y-intercept to be 3. Write the slope intercept form of the line as $y = -1x + 3$. Add x to both sides to obtain $x + y = 3$, the equation in standard form.

(Easy) (Skill 3.4)

109. **Identify the proper sequencing of subskills when teaching graphing inequalities in two dimensions.**

A. Shading regions, graphing lines, graphing points, determining whether a line is solid or broken

B. Graphing points, graphing lines, determining whether a line is solid or broken, shading regions

C. Graphing points, shading regions, determining whether a line is solid or broken, graphing lines

D. Graphing lines, determining whether a line is solid or broken, graphing points, shading regions

Answer: B. Graphing points, graphing lines, determining whether a line is solid or broken, shading regions

(Rigorous) (Skill 3.3)

110. **What is the equation of the below graph?**

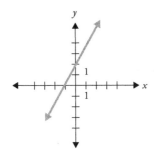

A. $2x + y = 2$

B. $2x - y = -2$

C. $2x - y = 2$

D. $2x + y = -2$

Answer: B. $2x - y = -2$

By observation, we see that the graph has a y-intercept of 2 and a slope of $\frac{2}{1} = 2$. Therefore its equation is $y = mx + b = 2x + 2$. Rearranging the terms gives $2x - y = -2$.

(Rigorous) (Skill 3.3)

111. **Which graph represents the equation of $y = x^2 + 3x$?**

A.

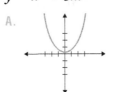

B.

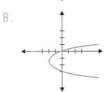

C.

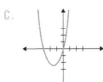

D.

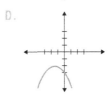

Answer: C.

B is not the graph of a function. D is the graph of a parabola where the coefficient of x^2 is negative. A appears to be the graph of $y = x^2$. To find the x-intercepts of $y = x^2 + 3x$, set $y = 0$ and solve for x: $0 = x^2 + 3x = x(x + 3)$ to get $x = 0$ or $x = -3$. Therefore, the graph of the function intersects the x-axis at $x = 0$ and $x = -3$.

(Rigorous) (Skill 1.20)

112. Solve for *x*.

$$3x^2 - 2 + 4(x^2 - 3) = 0$$

A. $\{-\sqrt{2}, \sqrt{2}\}$

B. $\{2, -2\}$

C. $\{0, \sqrt{3}, -\sqrt{3}\}$

D. $\{7, -7\}$

Answer: A. $\{-\sqrt{2}, \sqrt{2}\}$

Distribute and combine like terms to obtain $7x^2 - 14 = 0$. Add 14 to both sides, and then divide by 7. Since $x^2 = 2$, $x = \pm\sqrt{2}$

(Average) (Skill 1.14)

113. Which of the following is a factor of $6 + 48m^3$?

A. $(1 + 2m)$

B. $(1 - 8m)$

C. $(1 + m - 2m)$

D. $(1 - m + 2m)$

Answer: A. $(1 + 2m)$

Removing the common factor of 6 and then factoring the sum of two cubes gives $6 + 48m^3 = 6(1 + 8m^3) = 6(1 + 2m)(1 - 2m + (2m)^2)$.

(Average) (Skill 1.14)

114. Factor completely:

$$8(x - y) + a(y - x)$$

A. $(8 + a)(y - x)$

B. $(8 - a)(y - x)$

C. $(a - 8)(y - x)$

D. $(a - 8)(y + x)$

Answer: C. $(a - 8)(y - x)$

Glancing first at the solution choices, factor $(y - x)$ from each term. This leaves -8 from the first term and a from the second term: $(a - 8)(y - x)$

(Average) (Skill 1.14)

115. Which of the following is a factor of $k^3 - m^3$?

A. $k^2 + m^2$

B. $k + m$

C. $k^2 - m^2$

D. $k - m$

Answer: D. $k - m$

The complete factorization for a difference of cubes is $(k - m)(k^2 + mk + m^2)$.

(Rigorous) (Skill 1.19)

116. **What is the solution set for the following equations?**

$$3x + 2y = 12$$
$$12x + 8y = 15$$

A. All real numbers

B. $x = 4, y = 4$

C. $x = 2, y = -1$

D. $\varnothing$

Answer: D. $\varnothing$

Multiplying the top equation by -4 and adding results in the equation $0 = -33$. Since this is a false statement, the correct choice is the null set.

(Rigorous) (Skill 1.19)

117. **Solve for x and y:**

$$x = 3y + 7$$
$$7x + 5y = 23$$

A. $(-1, 4)$

B. $(4, -1)$

C. $\left(\frac{-29}{7}, \frac{-26}{7}\right)$

D. $(10, 1)$

Answer: B. $(4, -1)$

Substituting x in the second equation results in $7(3y + 7) + 5y = 23$. Solve by distributing and grouping like terms: $26y + 49 = 23$, $26y = -26$, $y = -1$. Substitute y into the first equation to obtain x.

(Rigorous) (Skill 1.19)

118. **Solve the system of equations for x, y, and z.**

$$3x + 2y - z = 0$$
$$2x + 5y = 8z$$
$$x + 3y + 2z = 7$$

A. $(-1, 2, 1)$

B. $(1, 2, -1)$

C. $(-3, 4, -1)$

D. $(0, 1, 2)$

Answer: A. $(-1, 2, 1)$

Multiplying equation 1 by 2, and equation 2 by -3, and then adding together the two resulting equations gives $-11y + 22z = 0$. Solving for y gives $y = 2z$. In the meantime, multiplying equation 3 by -2 and adding it to equation 2 gives $-y - 12z = -14$. Then substituting $2z$ for y, yields the result $z = 1$. Subsequently, one can easily find that $y = 2$, and $x = -1$.

(Rigorous) (Skill 1.20)

119. Find the zeroes of $f(x) = x^3 + x^2 - 14x - 24$.

   A. 4, 3, 2

   B. 3, -8

   C. 7, -2, -1

   D. 4, -3, -2

Answer: D. 4, -3, -2

Possible rational roots of the equation $0 = x^3 + x^2 - 14x - 24$ are all the positive and negative factors of 24. By substituting into the equation, we find that -2 is a root, and therefore that $x + 2$ is a factor. By performing the long division $\frac{(x^3 + x^2 - 14x - 24)}{(x + 2)}$, we can find that another factor of the original equation is $x^2 - x - 12$ or $(x - 4)(x + 3)$. Therefore the zeros of the original function are -2, -3, and 4.

(Rigorous) (Skill 1.20)

120. The discriminant of a quadratic equation is evaluated and determined to be -3. The equation has:

   A. One real root

   B. One complex root

   C. Two roots, both real

   D. Two roots, both complex

Answer: D. Two roots, both complex

The discriminant is the number under the radical sign. Since it is negative, the two roots of the equation are complex.

(Rigorous) (Skill 3.3)

121. Which equation is graphed below?

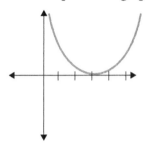

   A. $y = 4(x + 3)^2$

   B. $y = 4(x - 3)^2$

   C. $y = 3(x - 4)^2$

   D. $y = 3(x + 4)^2$

Answer: B. $y = 4(x - 3)^2$

Since the vertex of the parabola is three units to the right of the origin, we choose the solution where 3 is subtracted from x, and then the quantity is squared.

(Easy) (Skill 3.2)

122. Which set illustrates a function?

   A. {(0,1) (0,2) (0,3) (0,4)}

   B. {(3, 9) (-3, 9) (4, 16) (-4, 16)}

   C. {(1, 2) (2, 3) (3, 4) (1, 4)}

   D. {(2, 4) (3, 6) (4, 8) (4, 16)}

Answer: B. {(3, 9) (-3, 9) (4, 16) (-4, 16)}

Each number in the domain can only be matched with one number in the range. A is not a function because 0 is mapped to 4 different numbers in the range. In C, 1 is mapped to two different numbers. In D, 4 is also mapped to two different numbers.

(Rigorous) (Skill 3.7)

123. **Give the domain for the function over the set of real numbers:**

$$y = \frac{3x + 2}{2x^2 - 3}$$

A. All real numbers

B. All real numbers, $x \neq 0$

C. All real numbers, $x \neq -2$ or 3

D. All real numbers, $x \neq \frac{\pm\sqrt{6}}{2}$

Answer: D. All real numbers, $x \neq \frac{\pm\sqrt{6}}{2}$

Solve the denominator for 0. These values will be excluded from the domain.

$$2x^2 - 3 = 0$$
$$2x^2 = 3$$
$$x^2 = \frac{3}{2}$$
$$x = \sqrt{\frac{3}{2}} = \sqrt{\frac{3}{2}} \times \sqrt{\frac{2}{2}} = \frac{\pm\sqrt{6}}{2}$$

(Rigorous) (Skill 3.8)

124. **If y varies inversely as x and x is 4 when y is 6, what is the constant of variation?**

A. 2

B. 12

C. $\frac{3}{2}$

D. 24

Answer: D. 24

The constant of variation for an inverse proportion is xy.

(Rigorous) (Skill 3.8)

125. **If y varies directly as x and x is 2 when y is 6, what is x when y is 18?**

A. 3

B. 6

C. 26

D. 36

Answer: B. 6

$\frac{x}{y} = \frac{2}{6} = \frac{x}{18}$, Solve $36 = 6x$.

(Rigorous) (Skill 3.7)

126. **State the domain of the function**

$$f(x) = \frac{3x - 6}{x^2 - 25}$$

A. $x \neq 2$

B. $x \neq 5, -5$

C. $x \neq 2, -2$

D. $x \neq 5$

Answer: B. $x \neq 5, -5$

The values of 5 and -5 must be omitted from the domain of all real numbers because if x took on either of those values, the denominator of the fraction would have a value of 0, and therefore the fraction would be undefined.

(Average) (Skill 3.5)

127. The volume of water flowing through a pipe varies directly with the square of the radius of the pipe. If the water flows at a rate of 80 liters per minute through a pipe with a radius of 4 cm, at what rate would water flow through a pipe with a radius of 3 cm?

A. 45 liters per minute

B. 6.67 liters per minute

C. 60 liters per minute

D. 4.5 liters per minute

Answer: A. 45 liters per minute

Set up the direct variation: $\frac{V}{r^2} = \frac{V}{r^2}$. Substituting gives $\frac{80}{16} = \frac{V}{9}$. Solving for V gives 45 liters per minute.

(Average) (Skill 3.8)

128. Three less than four times a number is five times the sum of that number and 6. Which equation could be used to solve this problem?

A. $3 - 4n = 5(n + 6)$

B. $3 - 4n + 5n = 6$

C. $4n - 3 = 5n + 6$

D. $4n - 3 = 5(n + 6)$

Answer: D. $4n - 3 = 5(n + 6)$

Be sure to enclose the sum of the number and 6 in parentheses.

(Average) (Skill 4A.7)

129. Find the median of the following set of data: 14 3 7 6 11 20

A. 9

B. 8.5

C. 7

D. 11

Answer: A. 9

Place the numbers in ascending order: 3 6 7 11 14 20. Find the average of the middle two numbers $\frac{(7 + 11)}{2} = 9$.

(Average) (Skill 4A.7)

130. Compute the median for the following data set: {12, 19, 13, 16, 17, 14}

A. 14.5

B. 15.17

C. 15

D. 16

Answer: C. 15

Arrange the data in ascending order: 12, 13, and 14,16,17,19. The median is the middle value in a list with an odd number of entries. When there is an even number of entries, the median is the mean of the two center entries. Here the average of 14 and 16 is 15.

(Average) (Skill 4A.7)

131. **Corporate salaries are listed for several employees. Which would be the best measure of central tendency?**

$24,000	$24,000	$26,000
$28,000	$30,000	$120,000

A. Mean

B. Median

C. Mode

D. No difference

Answer: B. Median

The median provides the best measure of central tendency in this case where the mode is the lowest number and the mean would be disproportionately skewed by the outlier $120,000.

(Average) (Skill 4A.7)

132. **Half the students in a class scored 80% on an exam, most of the rest scored 85% except for one student who scored 10%. Which would be the best measure of central tendency for the test scores?**

A. Mean

B. Median

C. Mode

D. Either the median or the mode because they are equal

Answer: B. Median

In this set of data, the median would be the most representative measure of central tendency since the median is independent of extreme values. Because of the 10% outlier, the mean (average) would be disproportionately skewed. In this data set, it is true that the median

and the mode (number which occurs most often) are the same, but the median remains the best choice because of its special properties.

(Average) (Skill 4A.8)

133. **A student scored in the 87th percentile on a standardized test. Which would be the best interpretation of his score?**

A. Only 13% of the students who took the test scored higher

B. This student should be getting mostly Bs on his report card

C. This student performed below average on the test

D. This is the equivalent of missing 13 questions on a 100 question exam

Answer: A. Only 13% of the students who took the test scored higher

Percentile ranking tells how the student compared to the norm or the other students taking the test. It does not correspond to the percentage answered correctly, but can indicate how the student compared to the average student tested.

(Easy) (Skill 4A.2)

134. **Which statement is true about George's budget?**

    A. George spends the greatest portion of his income on food

    B. George spends twice as much on utilities as he does on his mortgage

    C. George spends twice as much on utilities as he does on food

    D. George spends the same amount on food and utilities as he does on mortgage

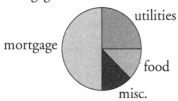

Answer: C. George spends twice as much on utilities as he does on food

(Average) (Skill 4A.2)

135. **What conclusion can be drawn from the graph below?**

    MLK Elementary Student Enrollment

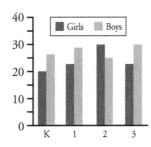

    A. The number of students in first grade exceeds the number in second grade.

    B. There are more boys than girls in the entire school

    C. There are more girls than boys in the first grade

    D. Third grade has the largest number of students

Answer: B. There are more boys than girls in the entire school

In Kindergarten, first grade, and third grade, there are more boys than girls. The number of extra girls in grade two is more than made up for by the extra boys in all the other grades put together.

(Average) (Skill 4A.3)

136. **Given a drawer with 5 black socks, 3 blue socks, and 2 red socks, what is the probability that you will draw two black socks in two draws in a dark room?**

    A. $\frac{2}{9}$

    B. $\frac{1}{4}$

    C. $\frac{17}{18}$

    D. $\frac{1}{18}$

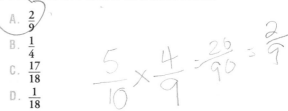

Answer: A. $\frac{2}{9}$

In this example of conditional probability, the probability of drawing a black sock on the first draw is $\frac{5}{10}$. It is implied in the problem that there is no replacement, therefore the probability of obtaining a black sock in the second draw is $\frac{4}{9}$. Multiply the two probabilities and reduce to lowest terms.

(Average) (Skill 4A.3)

137. **A sack of candy has 3 peppermints, 2 butterscotch drops, and 3 cinnamon drops. One candy is drawn and replaced, then another candy is drawn; what is the probability that both will be butterscotch?**

    A. $\frac{1}{2}$

    B. $\frac{1}{28}$

    C. $\frac{1}{4}$

    D. $\frac{1}{16}$

Answer: D. $\frac{1}{16}$

With replacement, the probability of obtaining a butterscotch on the first draw is $\frac{2}{8}$ and the probability of drawing a butterscotch on the second draw is also $\frac{2}{8}$. Multiply and reduce to lowest terms.

(Easy) (Skill 4A.3)

138. **Given a spinner with the numbers one through eight, what is the probability that you will spin an even number or a number greater than four?**

    A. $\frac{1}{4}$

    B. $\frac{1}{2}$

    C. $\frac{3}{4}$

    D. 1

Answer: C. $\frac{3}{4}$

There are 6 favorable outcomes: 2, 4, 5,6,7,8 and 8 possibilities. Reduce $\frac{6}{8}$ to $\frac{3}{4}$.

(Rigorous) (Skill 4A.3)

139. **If a horse will probably win three races out of ten, what are the odds that he will win?**

    A. 3:10

    B. 7:10

    C. 3:7

    D. 7:3

Answer: C. 3:7

There are 3 chances that the horse will win to 7 chances that he will not.

(Rigorous) (Skill 4A.3)

140. **How many ways are there to choose a potato and two green vegetables from a choice of three potatoes and seven green vegetables?**

    A. 126

    B. 63

    C. 21

    D. 252

Answer: B. 63

There are 3 ways to choose a potato and $7!/(5! \times 2!) = 21$ ways to choose two vegetables from a choice of seven. So the total number of ways one can choose a potato and two vegetables is $3 \times 21 = 63$.

(Average) (Skill 4B.2)

141. **Determine the number of subsets of set K. $K = \{4, 5, 6, 7\}$**

    A. 15

    B. 16

    C. 17

    D. 18

Answer: B. 16

A set of n objects has n^2 subsets. Therefore, here we have $4^2 = 16$ subsets. These subsets include four which each have 1 element only, six which each have 2 elements, four which each have 3 elements, plus the original set, and the empty set.

(Average) (Skill 4B.4)

142. $\{1, 4, 7, 10, \ldots\}$

What is the 40th term in this sequence?

A. 43

B. 121

C. 118

D. 120

Answer: C. 118

This ia an arithmetic series with the first term 1 and common difference 3. The 40th term $= 1 + (40 - 1)3 = 1 + 117 + 118$.

(Average) (Skill 4B.4)

143. $\{6, 11, 16, 21, \ldots\}$

Find the sum of the first 20 terms in the sequence.

A. 1070

B. 1176

C. 969

D. 1069

Answer: A. 1070

This ia an arithmetic series with the first term 6 and common difference 5. The sum of the first 20 terms $= (\frac{20}{2})(12 + (20 - 1)5) = 10\,(12 + 95) = 1070$.

(Rigorous) (Skill 4B.4)

144. **Find the sum of the first one hundred terms in the progression.** $(-6, -2, 2 \ldots)$

A. 19,200

B. 19,400

C. -604

D. 604

Answer: A. 19,200

To find the 100th term: $t_{100} = -6 + 99(4) = 390$. To find the sum of the first 100 terms: $S = \frac{100}{2}(-6 + 390) = 19200$.

(Rigorous) (Skill 4B.7)

145. **What would be the seventh term of the expanded binomial** $(2a + b)^8$?

A. $2ab^7$

B. $41a^4b^4$

C. $112a^2b^6$

D. $16ab^7$

Answer: C. $112a^2b^6$

The set-up for finding the seventh term is $\frac{8(7)(6)(5)(4)(3)}{6(5)(4)(3)(2)(1)}(2a)^{8-6}b^6$ which gives $28(4a^2b^6)$ or $112a^2b^6$.

(Easy)(Domain V)

146. **Which is the least appropriate strategy to emphasize when teaching problem solving?**

A. Guess and check

B. Look for key words to indicate operations such as all together—add, more than, subtract, times, multiply

C. Make a diagram

D. Solve a simpler version of the problem

Answer: B. Look for key words to indicate operations such as all together—add, more than, subtract, times, multiply

CPSIA information can be obtained at www.ICGtesting.com
Printed in the USA
LVOW02s0014160714

394456LV00029B/186/P